DOLL SEWING BOOK

HANON
娃衣缝纫书

（日）藤井里美（SATOMI FUJII）著
费军伟　张艳辉　译

化学工业出版社
·北京·

这是一本给娃娃做衣服的书，作品造型简约，但是注重版型和细节设计，每款衣服的布料、配色、配饰都由娃衣设计大师藤井里美亲自设计。她的娃衣品牌HANON更为知名，多次与多家知名娃娃品牌推出联名款设计，很多娃友对藤井里美的名字很陌生，但是对HANON耳熟能详。HANON的作品配色大胆独特，低调中显露高贵，面料多天然材质，制作步骤比较简单，易于互相搭配，具有自己独特的设计风格。书里的每款衣服都有S、M、L码，分别适合给中布娃娃、小布娃娃和荒木少女，也可以试试给其他尺码相近的娃娃穿着。书后附送裁剪用纸样，描下后裁剪，就可以反复使用，也可当作基础版，自行调整细节和尺寸。

北京市版权局著作权合同登记号：01-2019-1929

图书在版编目（CIP）数据

HANON娃衣缝纫书 /（日）藤井里美著；费军伟，张艳辉译. 一北京：化学工业出版社，2019.9（2025.7重印）
书名原文：DOLL SEWING BOOK HANON
ISBN 978-7-122-34428-1

Ⅰ. ①H… Ⅱ. ①藤… ②费… ③张… Ⅲ. ①手工艺品-制作
Ⅳ. ①TS973.5

中国版本图书馆CIP数据核字（2019）第085572号

责任编辑：高　雅　　装帧设计：王秋萍
责任校对：宋　夏

出版发行：化学工业出版社（北京市东城区青年湖南街 13 号　邮政编码 100011）
印　　装：北京宝隆世纪印刷有限公司
880mm×1092mm　1/16　印张 6　字数 160 千字　2025 年 7 月北京第 1 版第 5 次印刷

购书咨询：010-64518888　　售后服务：010-64518899
网　　址：http://www.cip.com.cn
凡购买本书，如有缺损质量问题，本社销售中心负责调换。

定　价：79.80 元

目录

这是一本关于娃娃的缝纫书。

造型简单，但追求理想剪裁。

缝纫初学者也可试着挑战。

大量使用的花边，

时尚流行的深色系，

可爱的糖果色系，

加上自己喜欢的精心装饰，

或者根据季节改变布料，体验创意乐趣。

S 码适用中布娃娃，

M 码适用小布娃娃，

L 码适用荒木少女，

根据各自特点制作衣服。

或许也适合其他尺码相近的娃娃，不妨试试。

This book is for making doll-size clothes.
The shape of the clothes is simple for an ideal silhouette.
The contents are made so that even a beginner can understand.

The styles can be modified for a wider range of looks
by using the colors and types of lace you prefer.
Please enjoy & have fun!

S size for Middie Blythe Doll
M size for Neo Blythe Doll
L size for Unoa 1.5 Girls

Those may suit the other dolls with the close size.
Please give it a try.

M 码缩褶绣连衣裙、围裙、蕾丝吊带连衣裙 & 鞋

M size Embroidered Smock Dress, Apron, Lace Strap Dress & Boots

EPINGLES

同款制作也很有趣的 M 码及 S 码。

并不是直接缩小版型，而是根据娃娃的实际身材，

逐个改变布料的用量、长度等。

M 码蕾丝吊带连衣裙 & 哈伦裤

M size Lace Strap Dress & Sarrouel Pants

M&S 码蕾丝吊带连衣裙、短裙 & 鞋

M & S size Lace Strap Dress, Skirt & Boots

M&S 码缩褶绣连衣裙 & 鞋

M & S size Embroidered Smock Dress & Boots

M 码彼得潘领连衣裙、袜子 & 俏皮的狐狸

M size Peter Pan Collar Dress, Socks & Sneaky Stuffed Fox

（上）S码外套、哈伦裤、单肩包 & 鞋

(上) *S size Coat, Sarrouel Pants, Shoulder Bag & Boots* (下) *S size Blouse, Vest, Trouser, Boots & Sneaky Stuffed Fox*

（下）S码罩衣、背心裤、鞋 & 俏皮的狐狸

M 码罩衣、短裙、袜子 & 鞋

M size Blouse, Skirt, Socks & Boots

M 码亚麻外套 & 鞋

M size Linen Arranged Coat & Boots

改变布料制作，整体印象焕然一新。
用皮革或丝带作为腰带，加上绣花装饰，
就能完成自己设计的款式。

实现S码及M码完美协调设计的"Tiny Betsy MaCall"娃娃，
可制作多款试穿。

S 码彼得潘领连衣裙、单肩包、M 码鞋 & 袜子

S size Peter Pan Collar Dress, Shoulder Bag, M size Boots & Socks

S 码缩褶绣连衣裙、哈伦裤 &M 码鞋

S size Embroidered Smock Dress, Sarrouel Pants & M size Boots

M码稍稍增加Betsy娃娃的剪裁长度。
对照身高，调节为合适长度。

M码蕾丝吊带连衣裙、胸花 & 俏皮的狐狸
M size Lace Strap Dress, Corsage & Sneaky Stuffed Fox

M 码短款缩褶绣连衣裙、M 码短裙 & 鞋

M size short-arranged Embroidered Smock Dress, M size Skirt & Boots

细长的1/6娃娃“UnoaQuluts Light”适合部分M尺码。
但是，衣长稍短，建议调节为合适长度。

M码缩褶绣连衣裙＆哈伦裤
M size Embroidered Smock Dress & Sarrouel Pants

M 码罩衣、背心、裤、袜子 & 鞋
M size Blouse, Vest, Trousers, Socks & Boots

L 码罩衣、背心、裤 & 鞋

L size Blouse, Vest, Trousers & Boots

L 尺码，两种尺寸胸围的荒木少女均适合穿着。
剪裁平缓的连衣裙等，还可穿在其他 40cm 身高娃娃的身上。

L 码缩褶绣连衣裙、蕾丝吊带连衣裙 & 哈伦裤

L size Embroidered Smock Dress, Lace Strap Dress & Sarrouel Pants

L 码蕾丝吊带连衣裙、哈伦裤 & 鞋

L size Lace Strap Dress, Sarrouel Pants & Boots

L 码外套、短裙 & 鞋

L size Coat, Skirt & Boots

L 码彼得潘领连衣裙、围裙、袜 & 鞋

L size Peter Pan Collar Dress, Apron, Socks & Boots

L 码缩褶绣连衣裙、蕾丝吊带连衣裙 & 鞋

L size Embroidered Smock Dress, Lace Strap Dress & Boots

设计宽松，外套和背心可叠穿。
还可尝试更多造型搭配。

L 码彼得潘领连衣裙、背心、胸花 & 单肩包
L size Peter Pan Collar Dress, Vest, Corsage & Shoulder Bag

工具
Tools

制作娃娃服之前，先要准备所需工具。
普通缝纫并不常用的工具，在制作小码娃娃服时也许能够派上大用场。

缎带 *Embroidery Silk Ribbon*
缎带绣花专用 3.5mm 宽缎带，柔软且使用方便，色数也很丰富。

绣花线 *Cotton Embroidery Floss*
基本上，使用 DMC 的 25 号线(单线)。

拆线刀 *Seam Ripper*
针脚不平时，可用其剪断线重新制作。

钳子 *Forceps*
将小布料翻到正面时手工用小钳子，非常方便。

线头剪 *Thread Scissors*
用于剪断手缝线或车缝线的线头。

顶针 *Thimble*
绣花或缭缝时使用。

锥子 *Embroidery Silk Ribbon*
布料翻面时钩出边角，或车缝时压住布料。

裁布剪 *Dressmaking Scissors*
选择裁剪整齐，方便细致操作的小型裁布剪。

缝线 *Sewing Thread*
车缝或手缝均使用 #90 号线。

布料胶 *Fabric Glue*
建议使用硬化后呈透明状的皮革、布料、纸等专用胶水。细微部位可使用方便的极细嘴。

锁边胶 *Fray Stopper*
裁剪之后，在布边涂抹锁边胶。

划粉笔 *Tailor's chalk*
薄布料选用不易渗漏的类型，灯芯绒等厚布料选用极细类型，深色布料选择白色划粉笔，区分使用。

蕾丝花边 *Laces*
花边太过鲜艳、难以搭配时，可进行草木染或红茶染之后调制成合适的颜色。

按扣 *Snaps*
使用 5mm 的圆按扣和 0 号弹簧扣。

缝针　珠针　大头针　直尺
Handsewing Needles, Dressmaker Pins, Silk Pin, Ruler

Lace Strap Dress

蕾丝吊带连衣裙

使用蕾丝花边制作的连衣裙，适合初学者。
肩带长度可调，可叠穿的百搭款。

全棉巴厘纱	S 30cm×20cm M 42cm×25cm L 93cm×45cm	裙子花边 （7～10mm宽）	S 13cm+4cm M 20cm+7cm L 45cm+12cm
5mm宽蕾丝	S 12cm×2根 M 14cm×2根 L 19cm×2根	蕾丝花边 按扣	覆盖衣片用量 S、M 1组 L 2组

1

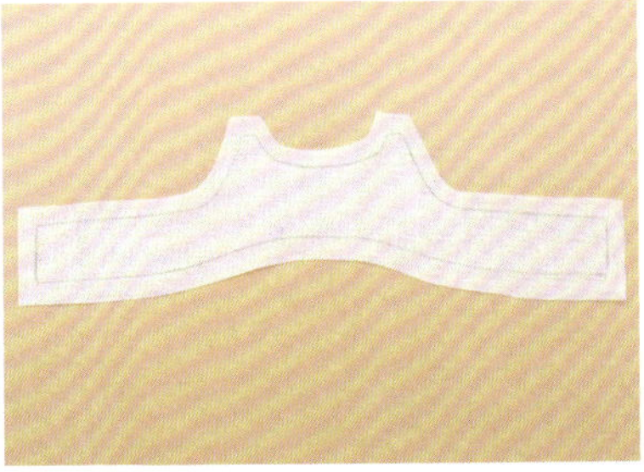

对照纸型裁剪各布件，布边涂抹锁边胶处理。

2

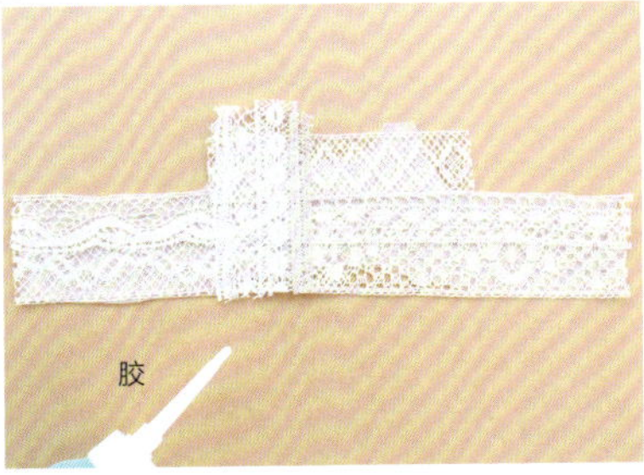
胶

碎花边放在正面衣片上，用布料胶预固定。

3

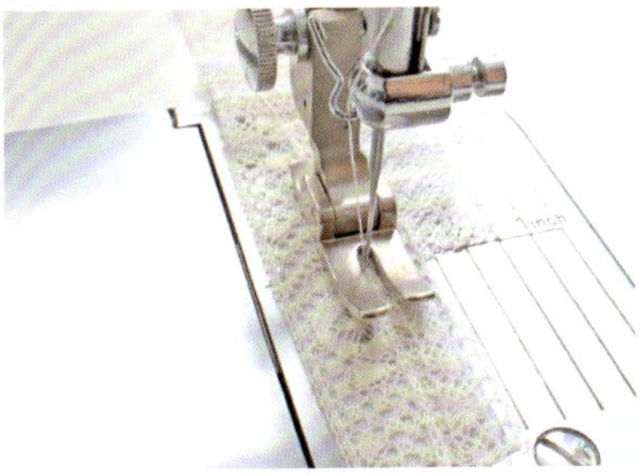

缝合花边端部，将花边缝接于衣片。

4

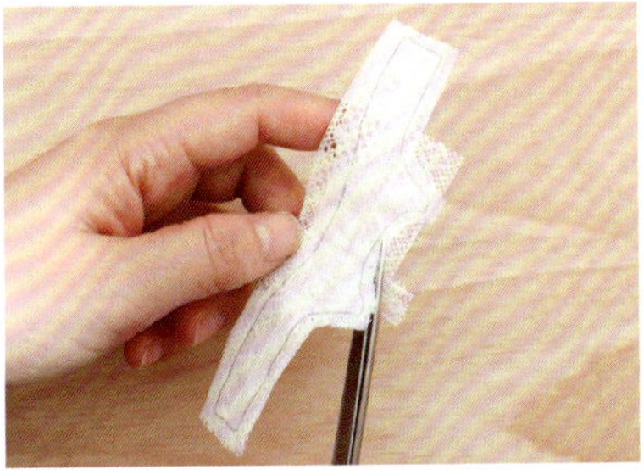

剪掉超出衣片的多余花边。（S）（M）为 7 号。

5

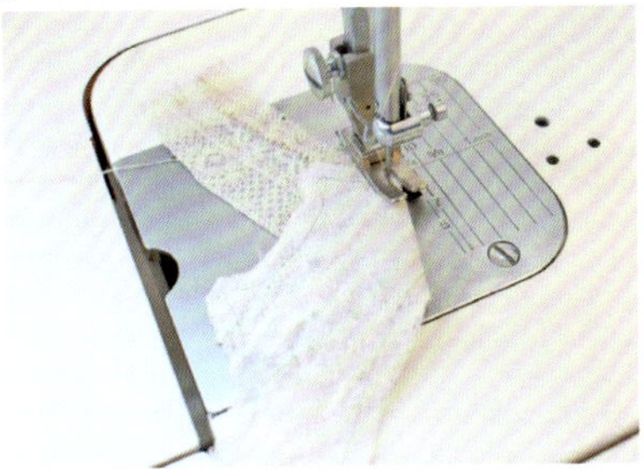

仅限（L）时，从衣片的两侧身正面向内对折衣褶缝合。

6

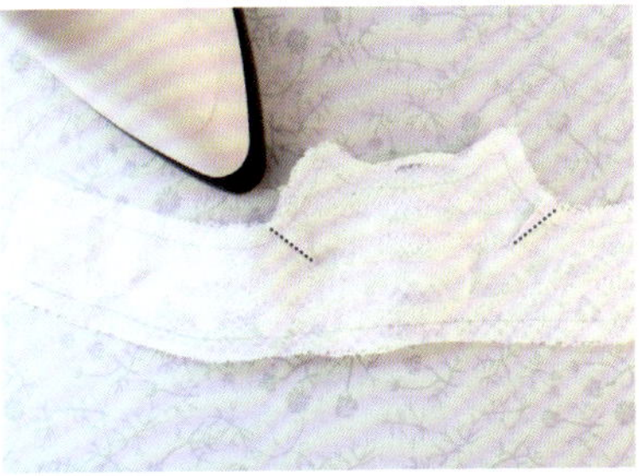

仅限（L）的衣褶压向内侧，并熨烫压实。

7

glue

在里衬衣片的肩部缝份侧涂抹少量布料胶，将 5mm 花边预固定于肩部。

8

正面对合衣片的表布及里衬，留下腰围部分，其余整周缝合。

9

剪掉缝份的边角，沿着弧线加入细密剪口。此时，注意避免剪到针脚。

This dress is a good place to start for beginners.
It is versatile and easy to coordinate.

{ materials : cotton voile, 7-10mm lace, 5mm lace, scrap lace, snaps }

1. Cut all parts. Put fray-stopper glue on the edges. *2.* Put lace on the front bodice with fabric glue. *3.* Stitch the edge of the lace.
4. Cut the extra lace. *5.* <L> size, Sew the front darts. *6.* <L>size, Iron the inside seam.
7. Glue shoulder lace onto the seam allowance of back bodice. *8.* Sew the bodice piece together inside out. *9.* Snip the seam allowance.

10

翻到正面，用锥子等将边角及弧线调整齐，并熨烫压实。

11

留下腰围部分，其余布边整周明线车缝压实。

12

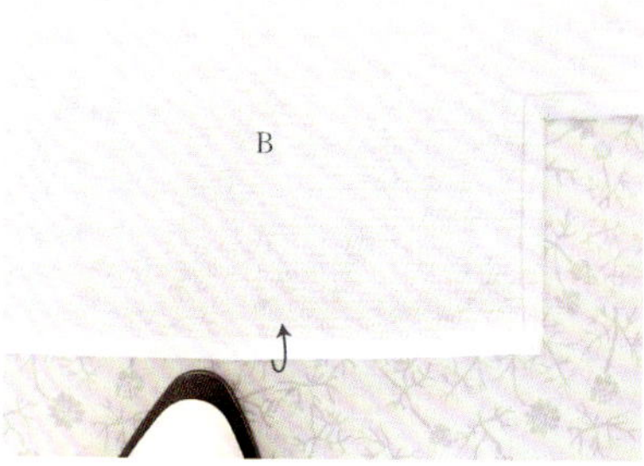
B

裙片 B 的底层裙摆的缝份折向内侧，缝合。

13

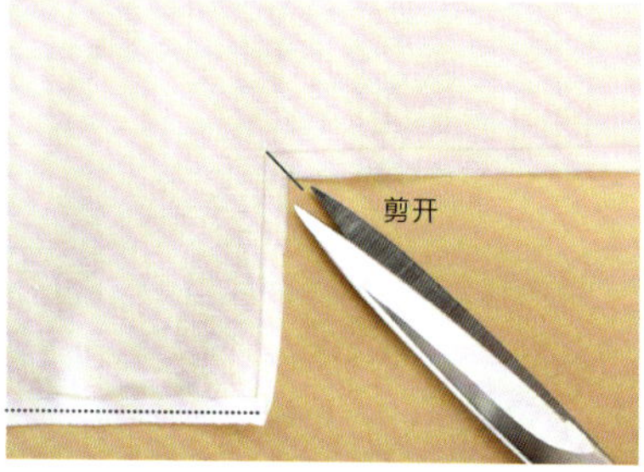
剪开

裙片 B 的上层裙摆的边角加入剪口。

14

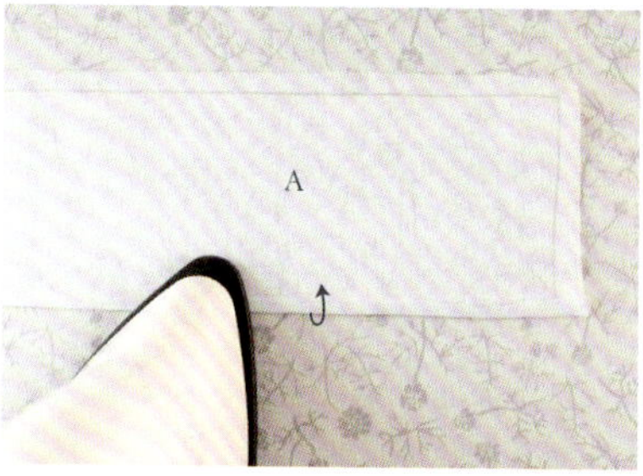
A

裙片 A 的裙摆缝份折向内侧，缝合。

15

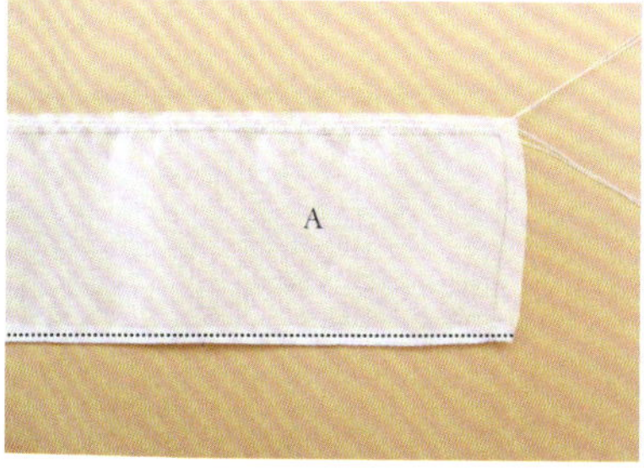
A

裙片上方缝份缩褶车缝。针脚调为约 2.5mm 宽度，缝份侧双线缝合。

16

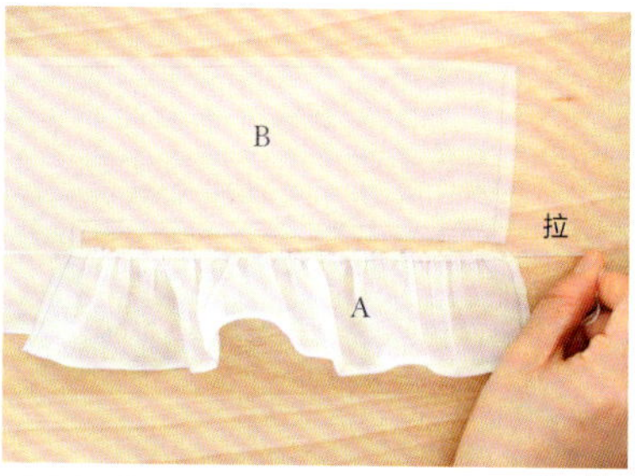
B
拉
A

对齐裙片 B 的上层裙摆（A 重叠部分）的宽度，缩褶（参照第 90 页）。

17

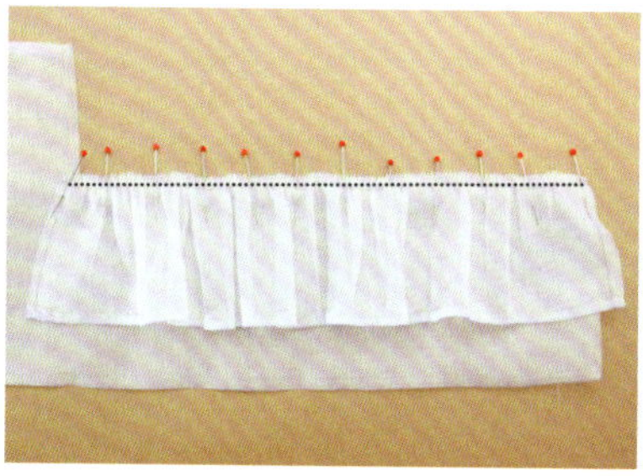

缩褶后的裙片 A 和 B 正面向对缝合。

18

缝份压向上方，熨烫压实。

10. Turn inside out and iron. *11.* Sew along the edge. *12.* Fold the lower hem of skirt B, iron and sew.
13. Snip the seam allowance. *14.* Fold the hem of skirt A, iron and sew. *15.* Gathering skirt A, seam allowance. [refer to P.90]
16. Pull the bobbin threads to match the width of fit the upper hem of skirt B. *17.* Sew skirt A to B face to face. *18.* Iron the seam allowance to face up.

19

裙片 A 和 B 的纵向部分同样正面向内缝合。缝份压向 B 侧，熨烫压实。

20

在裙片正面放上合适宽度的花边，用布料胶预固定，并缝合。

21

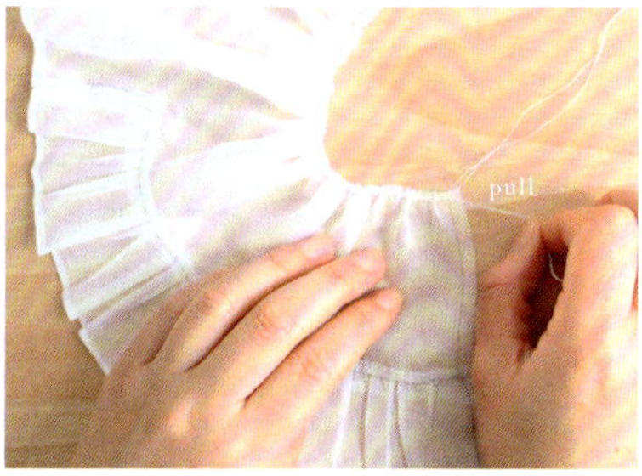
pull

裙片的腰围缝份侧，按照约 3mm 的针脚宽度双线缩褶车缝。

22

将裙片后中心的正品线和衣片的腰围宽度对齐进行缩褶，并正面向对缝合。

23

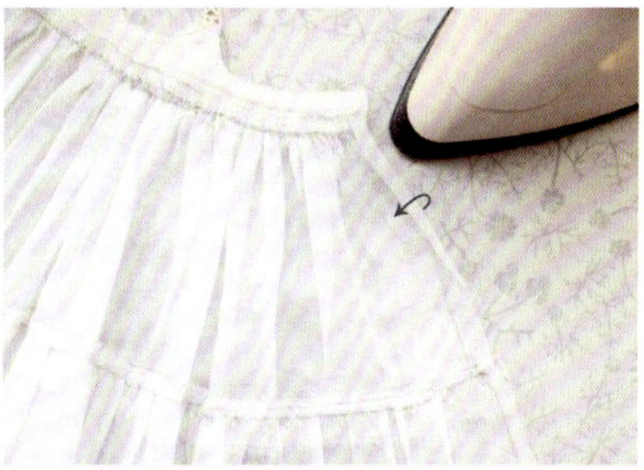

缝份压向衣片侧。接着，裙片的后开衩折向内侧（“开衩止处”拼合标记的正下方为止），熨烫压实。

24

腰围衣片侧明线车缝。

25

缝合倾斜折入的后开衩部分。

26

将裙片裙摆开始至“开衩止处”的后中心正面向对缝合。

27

熨烫摊开缝份，翻到正面缝接按扣之后即完成。用水弄湿后自然干燥，呈现做旧感。

19. Sew the sides of A and B together as shown. *20.* Put temporary and stitch lace on the seam of the skirt.
21. Gather the waist of the skirt until the width fits the bodice. *22.* Sew the waist of the bodice and skirt, inside out.
23. Iron the seam to bodice side. Fold the back opening. *24.* Sew the edge of the bodice from the front. *25.* Sew the back opening.
26. Sew the back and opening together inside out. *27.* Split open the seam allowance and iron down. Fasten snap at the back opening.

Sarrouel Pants

哈伦裤

容易搭配的款式。
使用暗色布料制作，更显帅气。

棉亚麻布	S 24cm×12cm M 26cm×15cm L 50cm×25cm	罗纹边用针织布	S 10cm×6cm M 14cm×7cm L 20cm×15cm
3mm 松紧绳	30cm	绣花线	原色、褐色

1

对照纸型裁剪各布件，布边涂抹锁边胶处理。针脚调整为约 2.5mm 宽度，双线缩褶车缝裤脚的缝份侧。

2

pull

对齐罗纹边宽度，缩褶（参照第 90 页）。

3

调整缩褶，熨烫压实。

4

将裤片的裤脚和罗纹边正面向对缝合。

5

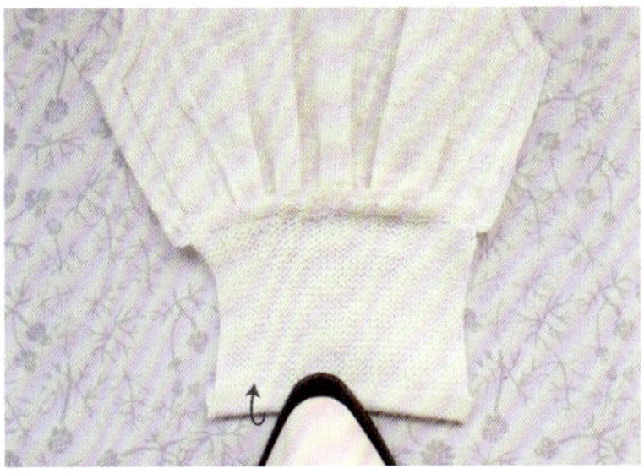

缝份压向上方，熨烫压实。罗纹边裤脚的缝份折向内侧，熨烫压实。

6

裤片侧明线车缝，罗纹边裤脚同样明线车缝。

7

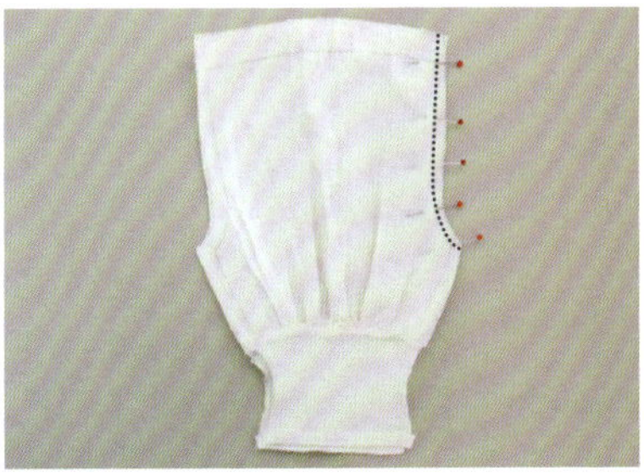

正面向对缝合裤片的前上裆。

8

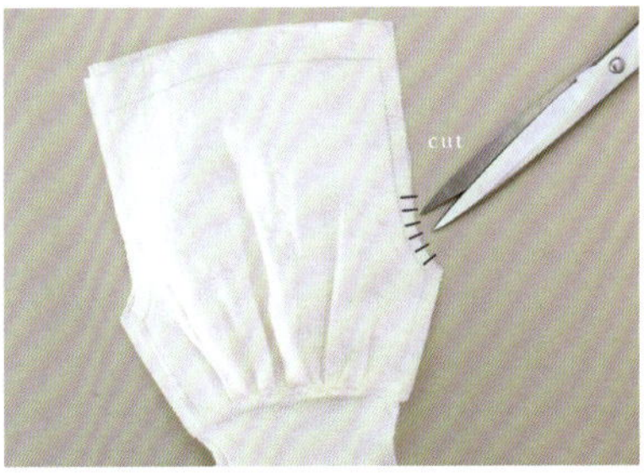
cut

前上裆的缝份加入剪口，熨烫摊开缝份。

9

三折边处理腰围的缝份，熨烫整齐。

Depending on the color you choose
these pants can be used in all kinds of outfit, dark for a chic look, light for natural.

{ materials : cotton linen, rib knit, 3mm elastic, embroidery thread }

1. Cut all parts. Put fray-stopper glue on the edges. Gather the pants hem seam allowance. [refer to P.90]
2. Pull the bobbin threads until the width the fits the elastic cuff. *3.* Iron the gathering.
4. Sew the pants to the cuff on the side. *5.* Iron the hem of the cuff up. *6.* Sew the top and bottom of the cuff hem.
7. Sew the front on the side as shown. *8.* Cut and iron the seam. *9.* Fold the waist seam down 3 times.

10

缝合腰围。

11

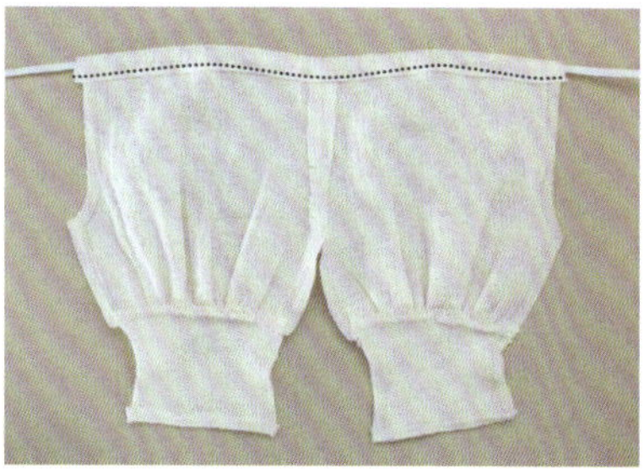

松紧绳穿入腰围。

12

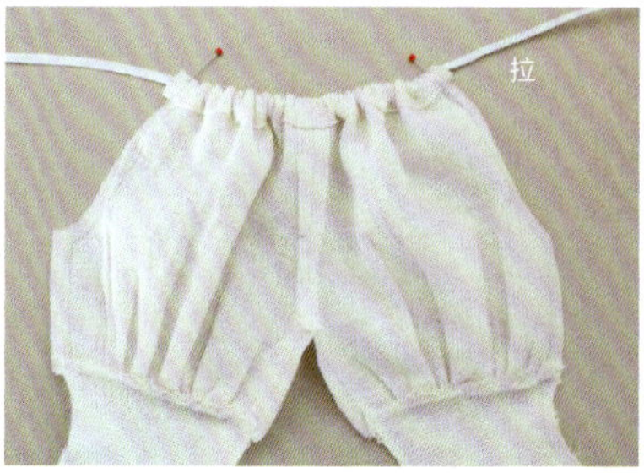
拉

收缩腰围宽度：（S）7.5cm，（M）9.5cm，（L）15cm，并用珠针固定。

13

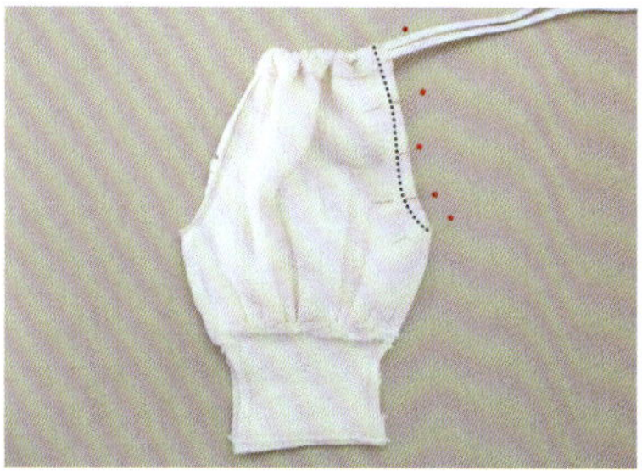

正面对合裤片的后上裆，连着松紧绳一起缝合。

14

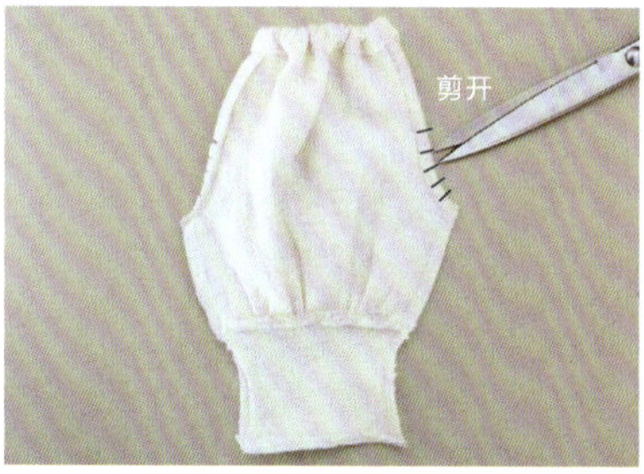
剪开

后上裆的缝份加入剪口，熨烫摊开缝份。

15

正面向对缝合下裆。

16

剪开

下裆的缝份加入剪口。

17

翻到正面，熨烫整齐之后，在上裆添加手缝装饰。

18

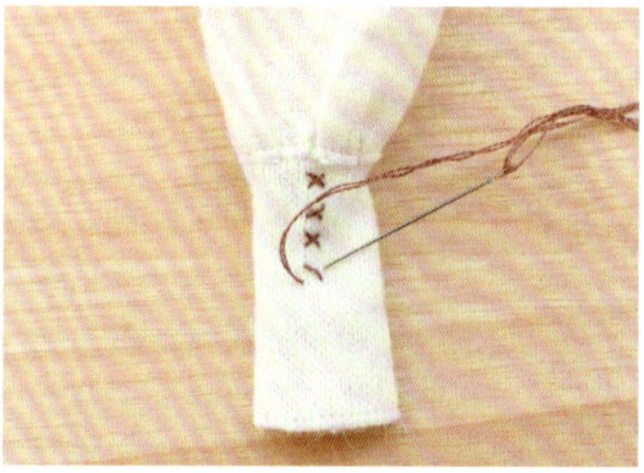

罗纹边的两侧边用褐色系绣花线交叉十字缝合。浸泡之后自然干燥，穿着舒适柔软。

10. Sew the folded. *11.* String elastic through the waist. *12.* Gather the waist to <S>7.5cm, <M>9.5cm or <L>15cm.
13. Sew the back together on the side as shown. *14.* Cut and iron the seam. *15.* Sew the inseam.
16. Cut the inseam. *17.* Turn the pants right side out. Hand sew the rise for decoration.
18. Use cross stitch to embroider the rib. Wash the pants in water and dry. This will give the cloth a natural finish.

Embroidered Smock Dress

缩褶绣连衣裙

接袖方式简单的合身连衣裙。
绣花也可根据个人喜好创意添加，或者缩短长度改成可爱的小罩衣。

亚麻布	S	30cm×30cm	罗纹边用针织布	S	2cm宽×11cm
	M	42cm×32cm		M	3cm宽×12cm
	L	90cm×55cm		L	3cm宽×20cm
按扣	S、M 1组 L 2组		绣花线	深蓝色或胭脂红	

1

将袖子的纸型描印于亚麻布料，稍大一圈裁剪之后固定于绣花框。

2

穿 1 根绣花线。首先，在中央双线回针绣（L 码为锁链绣）。

3

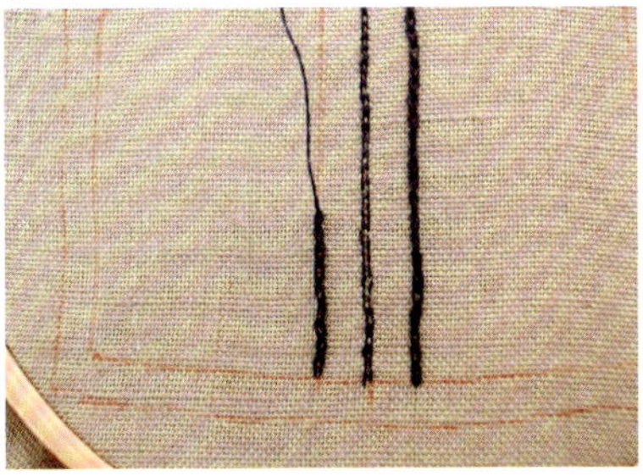

左右加入稍小的锁链绣。

4

在 3 个线条之间，加入法式结粒绣和十字绣。

5

中央、左右的线条加入平针绣，两端 V 字绣。绣花完成之后裁剪袖子，布边涂抹锁边胶处理。

6

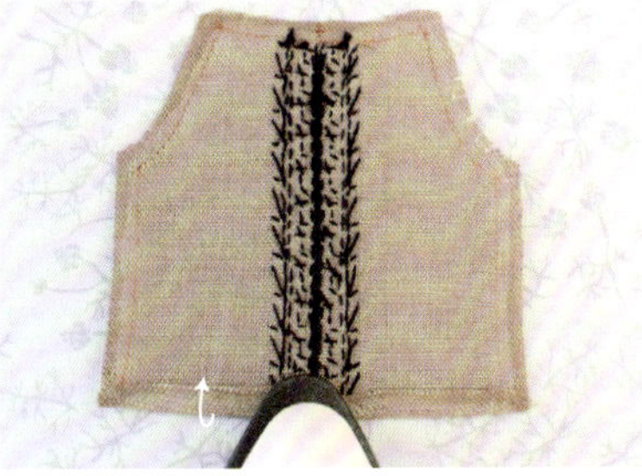

袖口的缝份折向内侧，熨烫压实。

7

拉

针脚调整为约 3mm 宽度，单线缩褶车缝袖口的缝份，拉线缩褶（参照第 90 页）。

8

按照（S）4cm、（M）5cm、（L）7.5cm 缩褶（不含缝份）之后打结，熨烫整齐之后明线车缝。

9

前衣片和袖子正面向对缝合。

This dress may look complicated but it is very simple to make.
The style of the embroidery is up to you, it can also be made as a blouse depending on the length.

{ materials : linen, cotton, snaps, embroidery thread }

1. Trace the sleeve pattern and stretch in an embroidery frame. *2.* Use one embroidery thread to back stitch the middle of the sleeves.
3. Use chain-stitch on either side. *4.* Use cross-stitch and French knot stitch alternatively.
5. Use running stitch along the chain and back stitch and V-stitch on outer side. Cut the sleeve and put fray-stopper glue on the edges.
6. Fold the edges of the sleeve opening and iron. *7.* Sew the edges of the sleeve opening 3mm stitch. [refer to P.90]
8. Gather the edges of the sleeve opening to <S> 3cm, <M> 5cm or <L> 7.5cm and sew. *9.* Sew the sleeves to the front.

10

接着，后衣片和袖子正面向内缝合。

11

剪开

按照约 5mm 间隔，在衣片和袖子的缝份加入剪口。

12

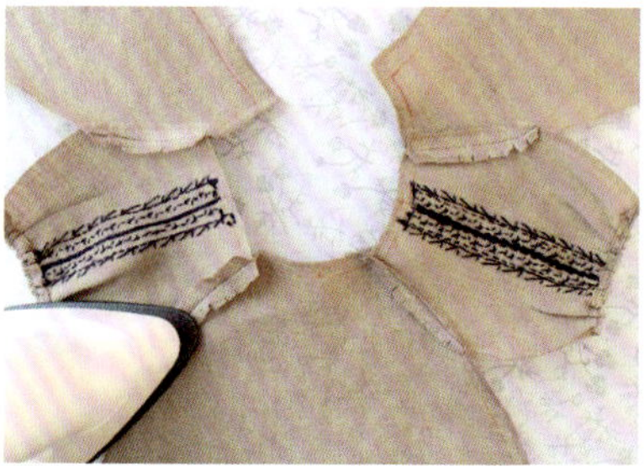

熨烫摊开缝份。

13

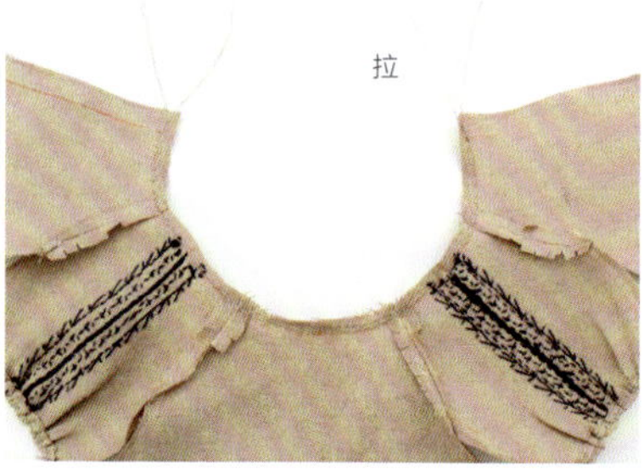
拉

针脚调整为约 3mm 宽度，双线缩褶车缝颈部周围的缝份（参照第 90 页）。

14

对齐颈部周围斜裁部分（※ 沿着布纹 45° 裁剪或使用斜裁布带）的宽度进行缩褶，熨烫整齐。

15

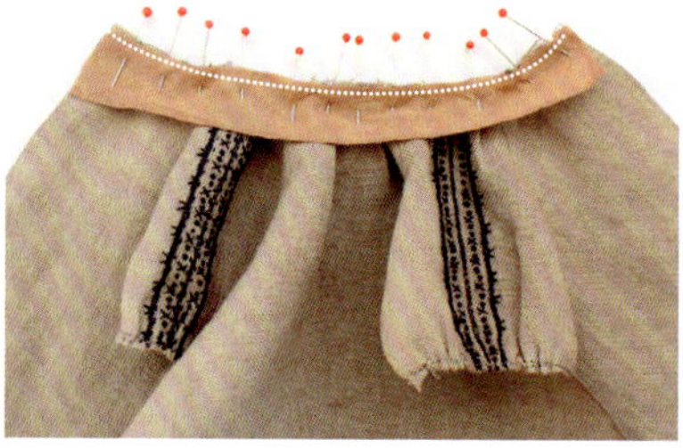

衣片的颈部周围和颈部斜裁布正面向内缝合。

16

缝份压向上方，熨烫压实。

17

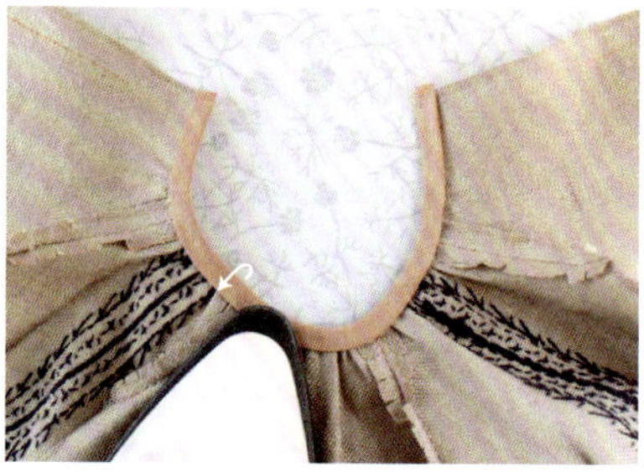

熨烫三折边斜裁部分，包住颈部周围的缝份。

18

缭缝固定斜裁部分的底边。

10. Sew the back bodice to the sleeves. *11.* Cut the seam. *12.* Unfold the seam and iron.
13. Sew 2 linens on the seam of the neck in 3mm stitch. [refer to P.90] *14.* Gather the seam of the neck opening until the width fits the bias tape.
15. Sew the bias tape to the neck. *16.* Iron the seam up. *17.* Fold the bias around the gather and iron. *18.* Finish with a blind stitch.

19

缝合完成。

20

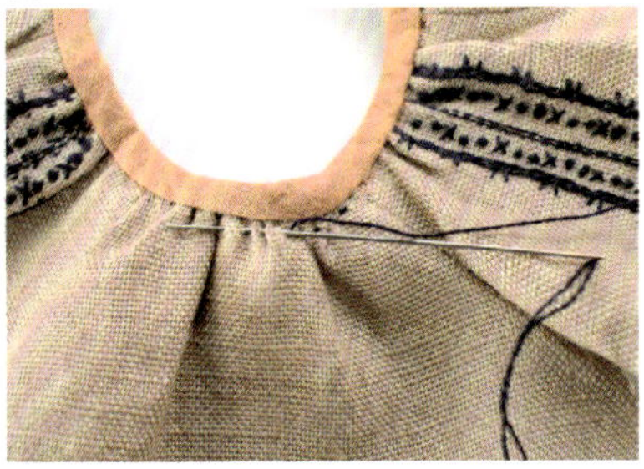

穿 1 根绣花线，在前衣片的颈部加入平针绣。

21

正面对合前衣片和后衣片，袖口、侧身、裙摆正面向对缝合。

22

剪开

侧身加入剪口。翻到正面，熨烫摊开缝份。

23

衣片裙摆的缝份折向内侧，熨烫压实。

24

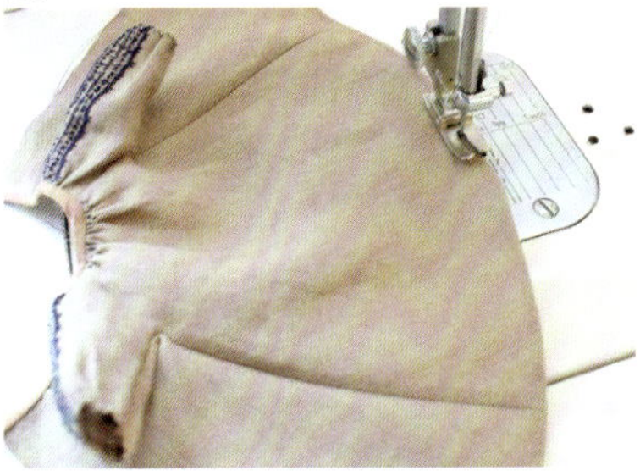

缝合裙摆。

25

后开衩折向内侧（至“开衩止处”拼合标记的下方附近），熨烫压实之后缝合。

26

从裙摆至“开衩止处”的后中心，正面向内缝合，缝份熨烫摊开。

27

翻到正面，缝接按扣之后完成。浸泡之后轻轻拧动并自然干燥，使布料自然柔软。

19. Now your neck opening is done! *20.* Use one embroidery thread to do a running-stitch on the middle of the front.
21. Sew the side of the bodice and the sleeves on the inside out. *22.* Make cuts in seam of the pits. Turn right side out and unfold the seam.
23. Fold the hem and iron. *24.* Sew the hem. *25.* Sew the front and back together inside out. *26.* Sew the back opening together.
27. Turn right side out. Fasten snap at the back opening. Wash the dress in water and dry. This will give the cloth a natural finish.

Apron
围裙

推荐使用越穿越有格调的亚麻布料。
简洁、自然的必备款式。

亚麻 S 42cm×10cm
M 44cm×14cm
L 73cm×25cm

1

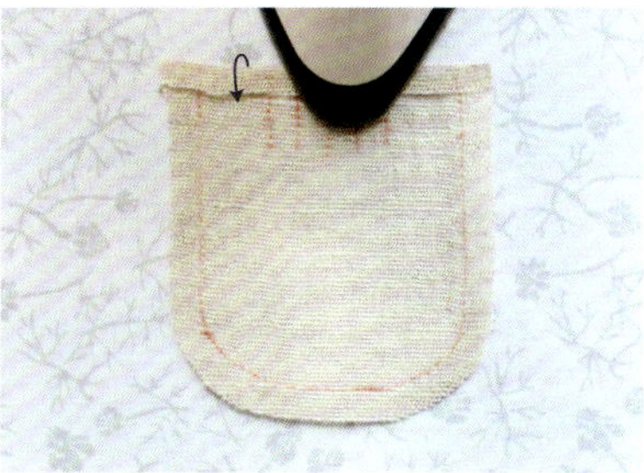

对照纸型裁剪各布件，布边涂抹锁边胶处理。折入袋口的缝份，熨烫压实。

2

折入

折叠细褶，用布料胶预固定。

3

缝合袋口。

4

针脚调整为约 3mm 宽度，拱缝缝合口袋弧线部分的缝份。

5

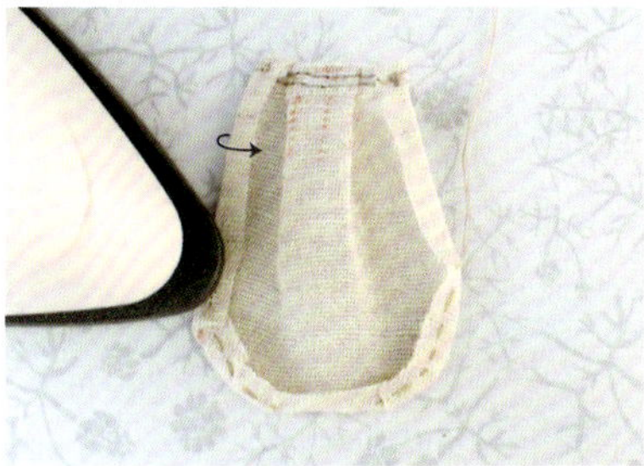

拉动拱缝的线进行缩褶，制作弧线之后熨烫整齐。

6

胶

缝份涂上布料胶，预固定于口袋位置。

7

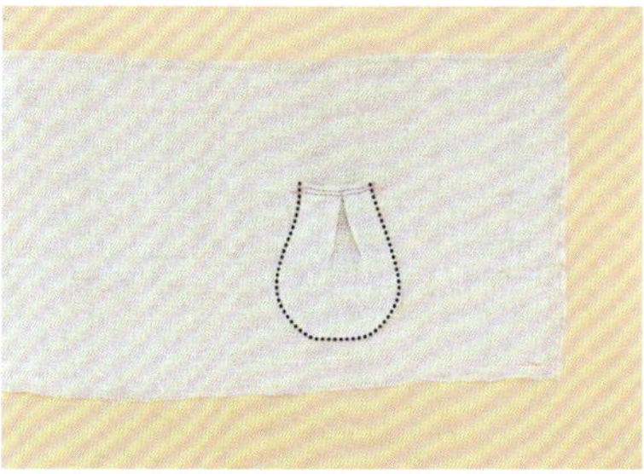

缝接口袋。

8

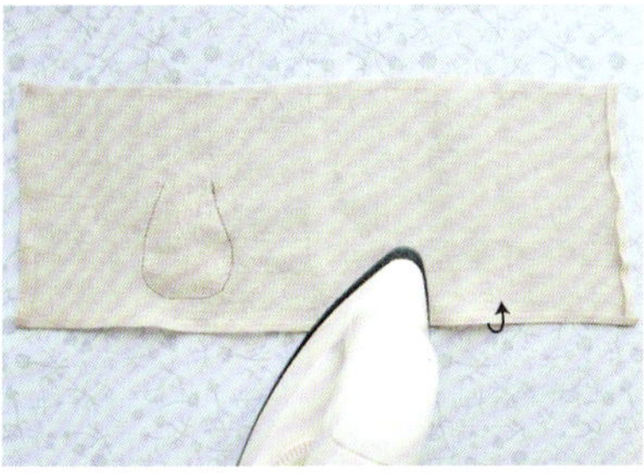

将罩衣的下摆及两端的缝份折向内侧，熨烫压实。

9

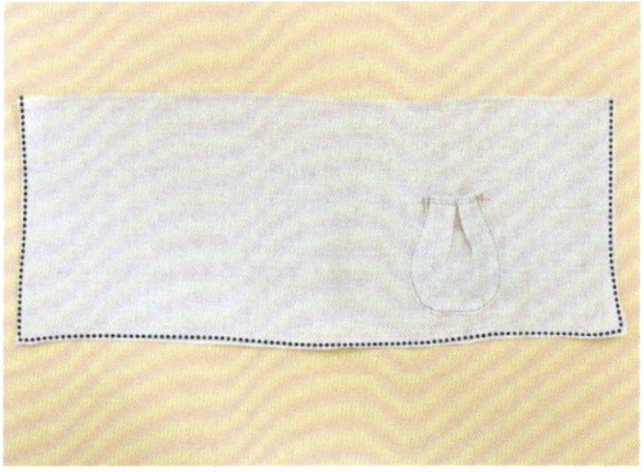

缝合端部及下摆。

For realism and a more worn look, linen is recommended.
A simple item like this can be used to complete any style of outfit.

{ material : linen }

1. Fold the top of the pocket and iron. *2.* Fold the tuck and glue with fabric glue. *3.* Sew the top of the pocket.
4. Sew the bottom of the pocket with running stitch and iron. *5.* Gather the bottom of the pocket and iron the seam.
6. Put small dots of fabric glue on the back of the pocket. *7.* Stick the pocket in place and sew. *8.* Fold and iron the hems as shown. *9.* Sew.

10

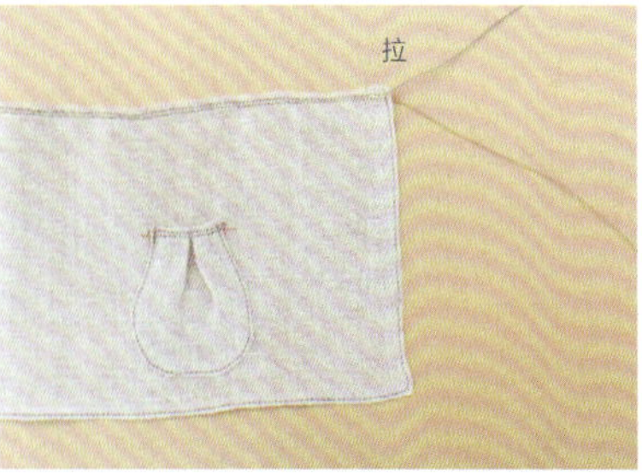
拉

针脚调整为约 3mm 宽度，双线缩褶车缝罩衣腰围的缝份（参照第 90 页）。

11

按照（S）7cm、（M）10cm、（L）14.5cm 缩褶，熨烫整齐。

12

正面对合罩衣和绳带，用珠针从中央开始均匀固定。

13

缝合罩衣和绳带。

14

缝份压向上方。

15

翻折绳带，熨烫压实。

16

三折边熨烫整齐绳带，包住缝份。

17

绳带的腰围位置明线车缝。

18

浸泡之后轻轻拧动并自然干燥，使布料自然柔软。

10. Sew 2 lines on the waist of the apron. [refer to P.90] *11.* Pull the bobbin threads until the width is <S>7cm, <M>10cm or <L>14.5cm and iron. *12.* Pin the waist cord to the apron. *13.* Sew the waist. *14.* Iron the seam up. *15.* Fold the waist cord once and iron. *16.* Fold the seam around the waist cord and iron. *17.* Sew the waist cord. *18.* Wash the apron in water and dry. This will give the cloth a natural finish.

Peter Pan Collar Dress

彼得潘领连衣裙

小彼得潘领、装饰围兜、袖头的颜色统一，与碎花布拼接搭配而成的精致连衣裙。
细褶难以处理的话可以放宽，缝上细花边同样漂亮。

花纹薄棉布	S	30cm×20cm	4mm 花边	S	14cm
	M	50cm×20cm		M	19cm
	L	105cm×25cm		L	35cm
纯色薄棉布	S	20cm×10cm	按扣	S、M 2组　L 3组	
	M	30cm×10cm	绣花线	珊瑚色	
	L	40cm×15cm			

1

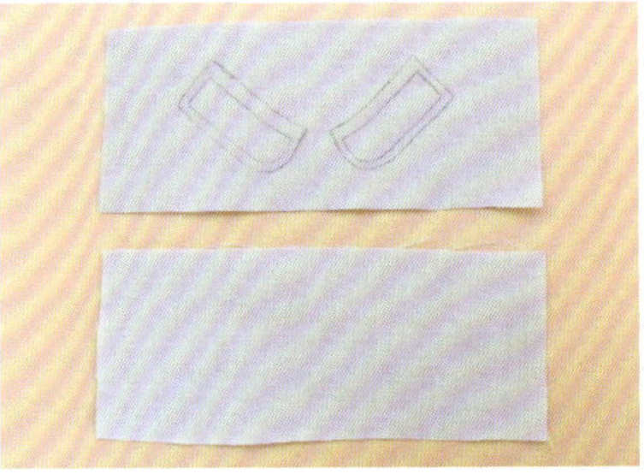

领子一对描印于布料，稍大一圈裁剪，准备另一片相同尺寸的布料。

2

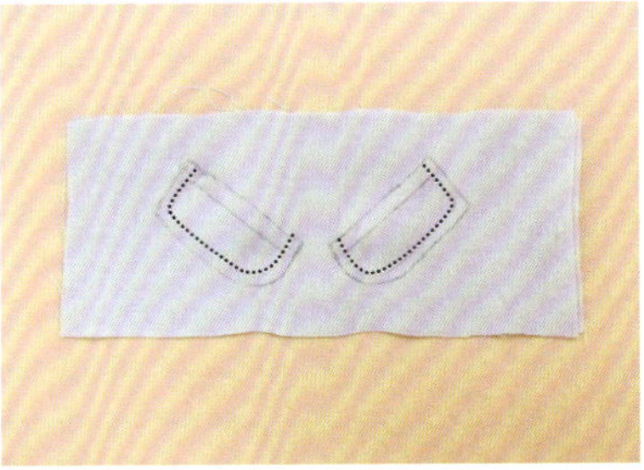

两片布料正面向对重合，缝合外侧的成品线。

3

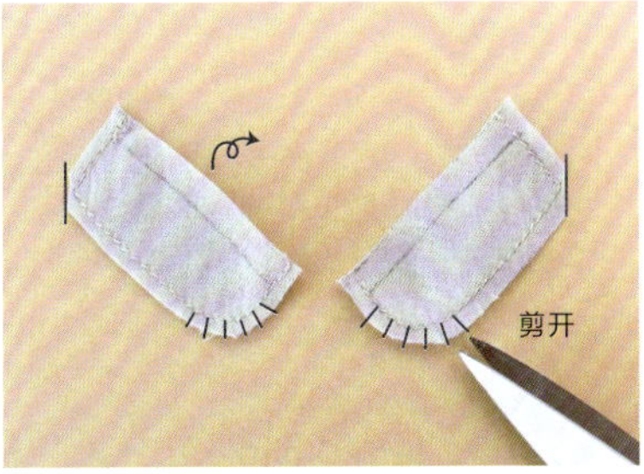
剪开

留下缝份裁剪领子，剪掉边角，沿着弧线加入细密剪口。此时，注意避免剪到针脚。

4

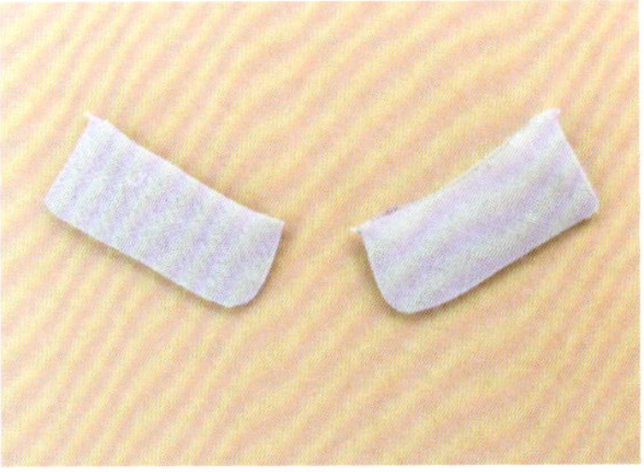

翻到正面，用锥子或钳子等将边角及弧线部分整齐翻出，熨烫整齐。

5

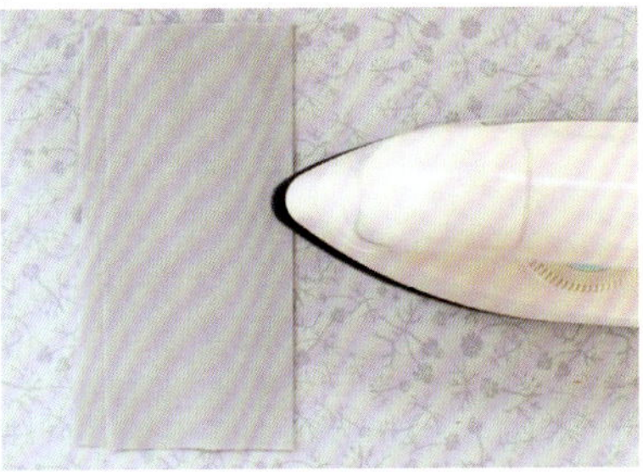

制作装饰围兜的细褶。沿着布纹，熨烫折入比装饰围兜纸型稍大一圈裁剪的布料。

6

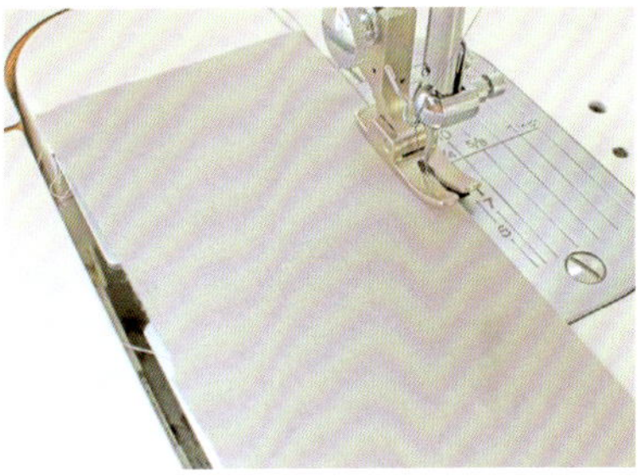

距离折痕 1mm 位置直线车缝。

7

缝合整齐的状态。

8

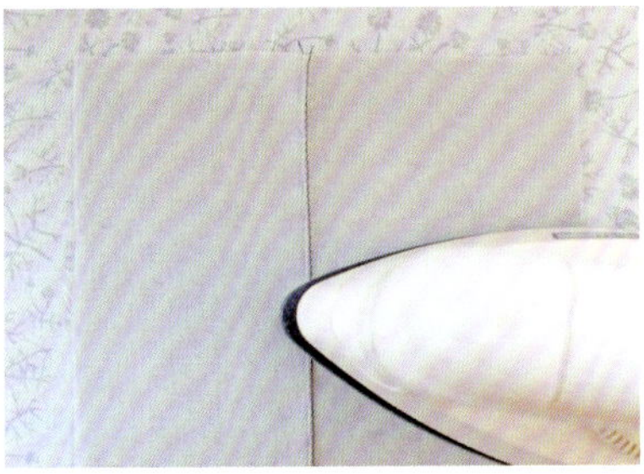

展开布料，将之前缝合的 1mm 宽度的折痕压向外侧，熨烫整齐。

9

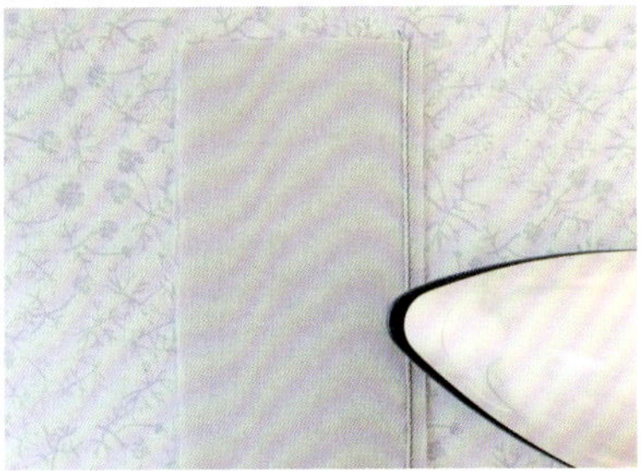

接着，在另一侧制作细褶。熨烫折入距离第 1 个折痕 6mm 位置。

Please have fun matching the colors of this design.
The pin-tucking can also be replaced by lace to make it easier.

{ materials : cotton/pattern fabric, 4mm lace, cotton/plain, snaps, embroidery thread }

1. For the collar take two pieces of the same size and draw the collar on one piece.
2. Take the collar pieces and match the edges, sew the blank lines as shown. *3.* Cut the shape of the collar and cut small cuts in the seam on the round.
4. Turn the collar pieces inside out and iron. *5.* Fold the piece of fabric for the bib. *6.* Sew the first seam 1mm from the edge.
7. The first line is done. *8.* Open the fold and iron. *9.* Fold the fabric under the first line with a 6mm space from the edge and iron.

10

车缝直线缝合距离折痕 1mm 位置。

11

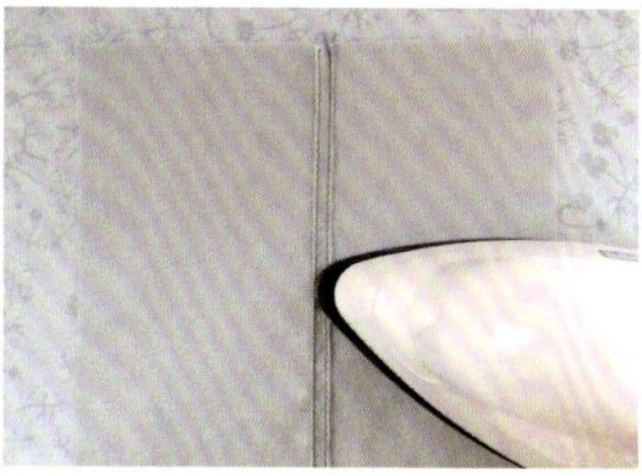

展开布料，折痕压向另一侧展开，熨烫整齐。

12

熨烫折入距离之前折痕 3mm 外侧位置，缝合距离折痕 1mm 位置，压向外侧。重复此操作，缝制细褶。

13

对齐细褶和中心，描印装饰围兜的纸型，并裁剪。其他各布件同样裁剪，布边涂抹锁边胶处理。

14

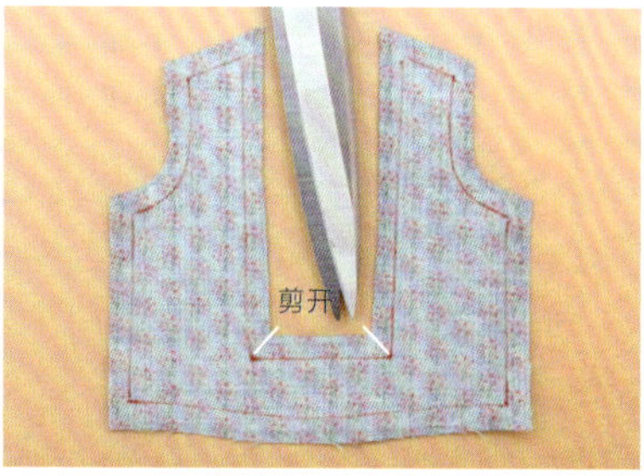
剪开

前衣片的缝份边角加入剪口。

15

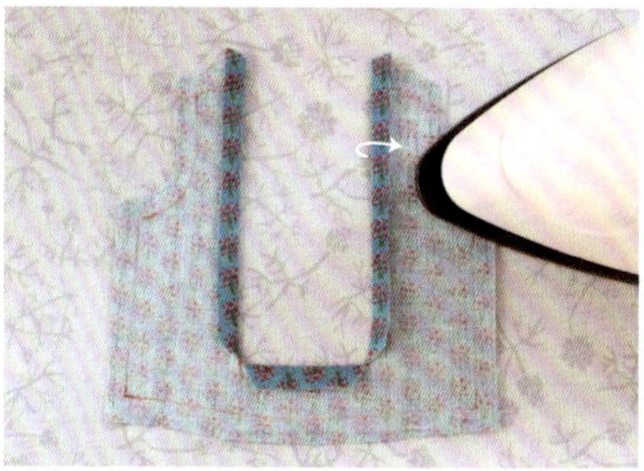

缝份折向内侧，熨烫压实。

16

用布料胶预固定前衣片和装饰围兜，车缝缝合。

17

花边放在边缘，用布料胶预固定，并缝接。

18

正面对合前衣片和后衣片，缝合肩部。

10. Sew 1mm in from the edge of the new fold. *11.* Open the fold and iron the new seam as shown.
12. Fold under 3mm from the new edge. Sew 1mm in from the edge. Repeat.
13. Trace the pattern for the bib on the pin tuck fabric. Put fray-stopper glue on the edges.
14. Cut the corners. *15.* Fold the seam and iron. *16.* Match the bib to the front with fabric glue and sew.
17. Put the lace on the front with fabric glue and sew. *18.* Match the front and back by the shoulders and pin, sew together.

19

熨烫摊开缝份。

20

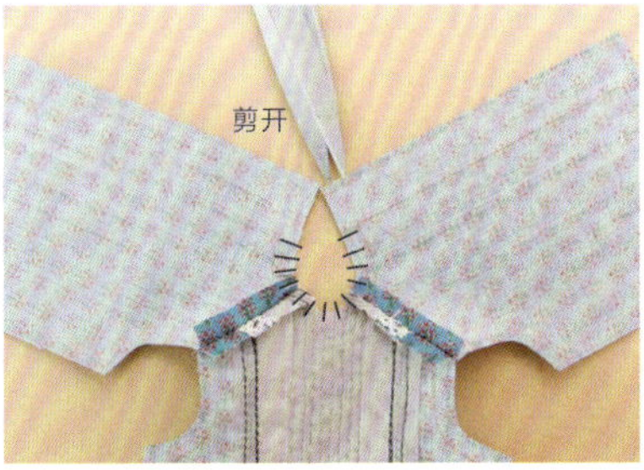
剪开

颈部周围的缝份加入细密剪口。

21

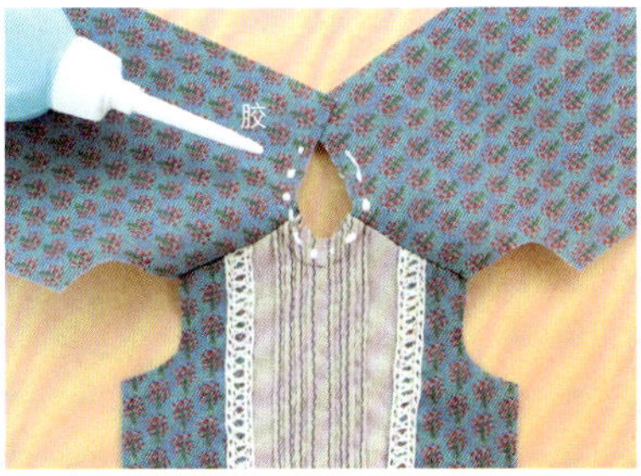
胶

缝份涂抹布料胶。

22

从装饰围兜的中心向左右均匀分配，预固定领子，并缝接。

23

剪开

领子的缝份加入细密剪口。

24

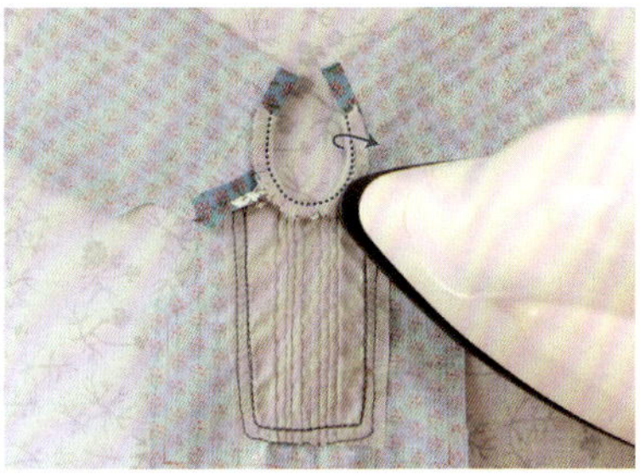

领子的缝份折向里衬，熨烫整齐之后，在颈部周围明线车缝。

25

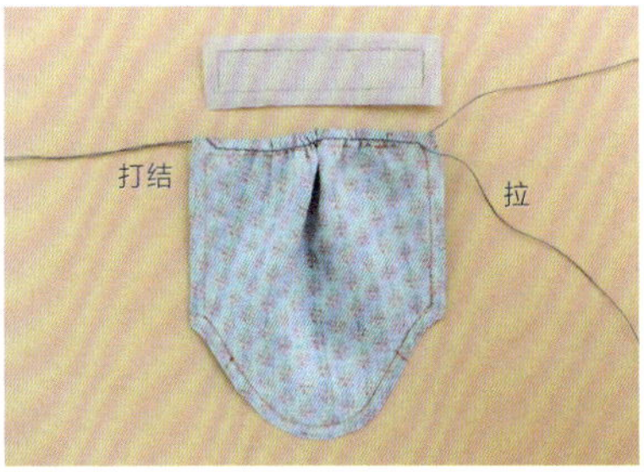
打结
拉

针脚调整为约 2.5mm 宽度，单线缩褶车缝袖口的缝份，对齐袖头的宽度进行缩褶（参照第 90 页）。

26

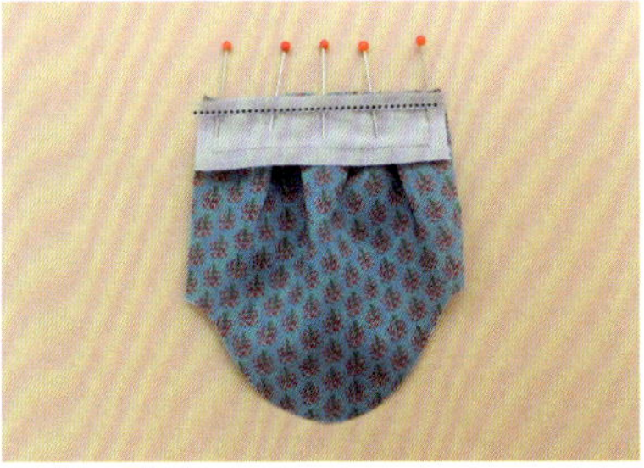

熨烫整齐缩褶部分，正面向内缝合袖头和袖口。

27

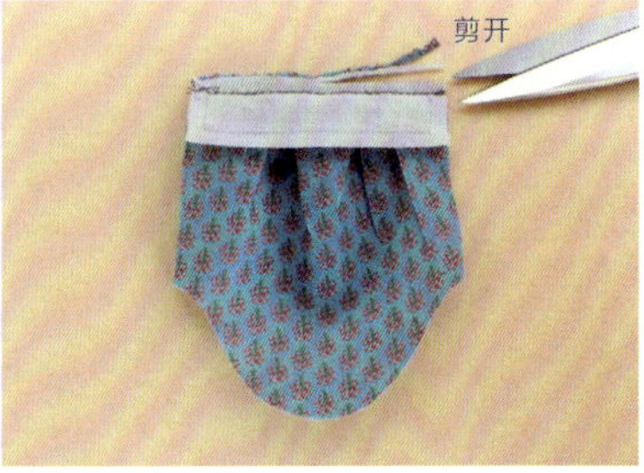
剪开

袖子和袖头的缝份裁剪为 3mm 宽度。

19. Unfold the seam and iron. *20.* Make small cuts on the seam of the neck opening. *21.* Put fabric glue on the edges.
22. Match the collar to the neck opening and sew. *23.* Make small cuts on the seam. *24.* Fold in side the seam, iron and sew the neck opening.
25. Gather the sleeve opening until the width fits the cuff. [refer to P.90] *26.* Match the sleeve openings and the cuff and sew. *27.* Cut the seam to 3mm.

28

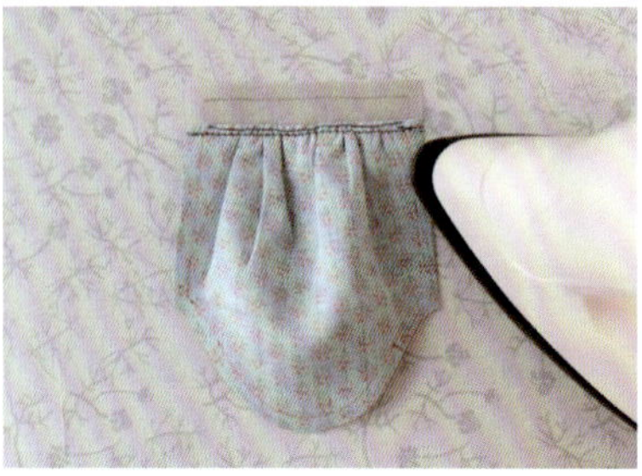

熨烫三折边袖头，包住缝份。

29

缭缝固定袖头的底边。

30

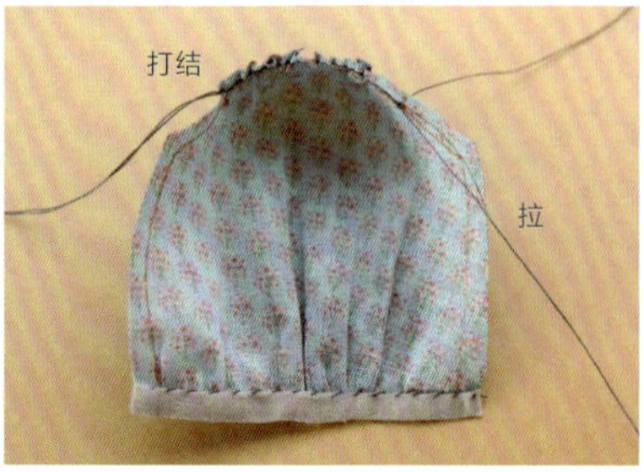
打结
拉

针脚调整为约 2.5mm 宽度，从袖山的缝份的拼合标记至拼合标记单线缩褶车缝。对齐衣片的袖窿宽度，进行缩褶。

31

正面向内缝合衣片和袖子。稍稍对合袖山的缝份和袖窿的缝份，多次车缝压实。

32

袖子已缝接于衣片。缝份压向袖子侧，熨烫整齐。

33

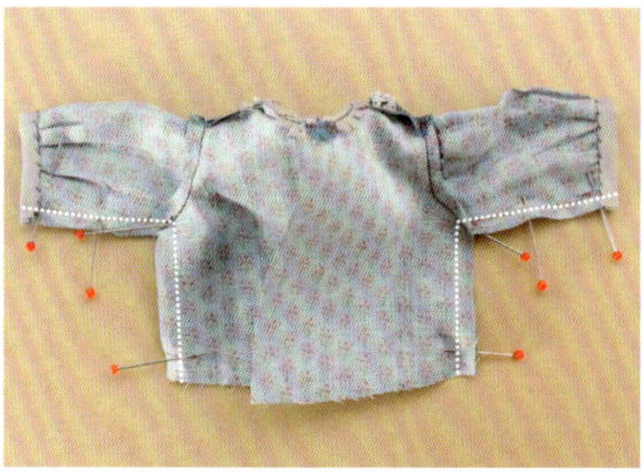

正面对合前衣片和后衣片，缝合袖口、侧身、裙片。

34

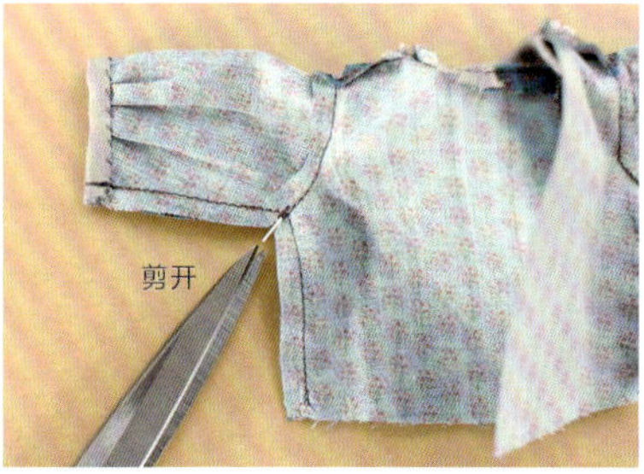
剪开

侧身的缝份加入剪口。翻到正面，熨烫摊开缝份。

35

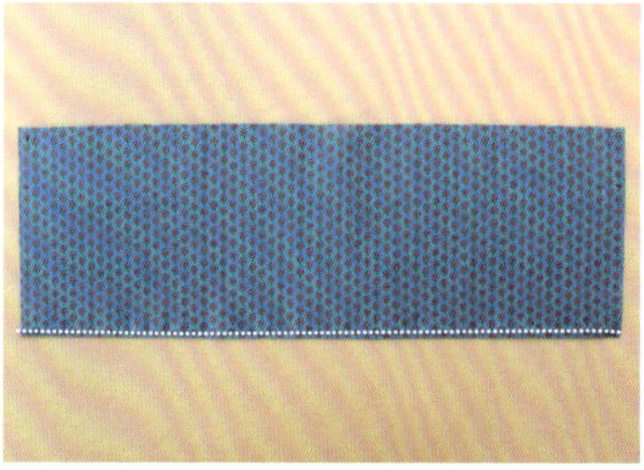

将裙摆的缝份熨烫折向内侧，缝合。

36

拉

针脚调整为约 2.5mm 宽度，双线缩褶车缝腰围的缝份。对齐衣片的腰围宽度，进行缩褶（参照第 90 页）。

28. Fold the cuff around the gathered sleeve. *29.* Finish with a blind stitch. *30.* Gather the shoulders until the width fits the armhole.
31. Match the side edge of the sleeve to the bodice and gradually sew the shoulder of the sleeve to the armhole, matching as you go.
32. Now your sleeves are attached. *33.* Pin the side of the bodice and sleeves and sew.
34. Make cuts in seam allowance of the pits. Turn right side out and unfold the seam, iron.
35. Fold the skirt hem, iron and sew. *36.* Gather the waist of the skirt until the width fits the waist. [refer P.90]

37

正面向内缝合衣片和裙片的腰围。

38

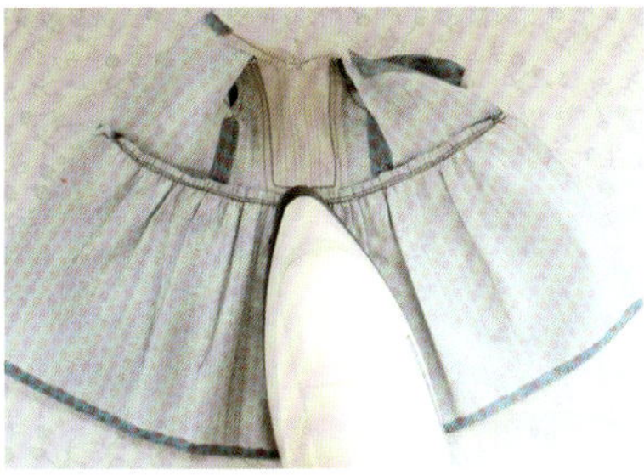

缝份压向衣片侧，熨烫压实。

39

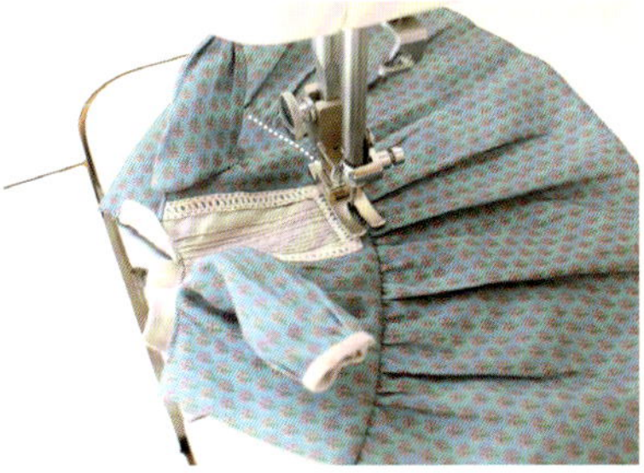

腰围衣片侧明线车缝。

40

后开衩折向内侧（“开衩止处”拼合标记的正下方为止），熨烫压实。

41

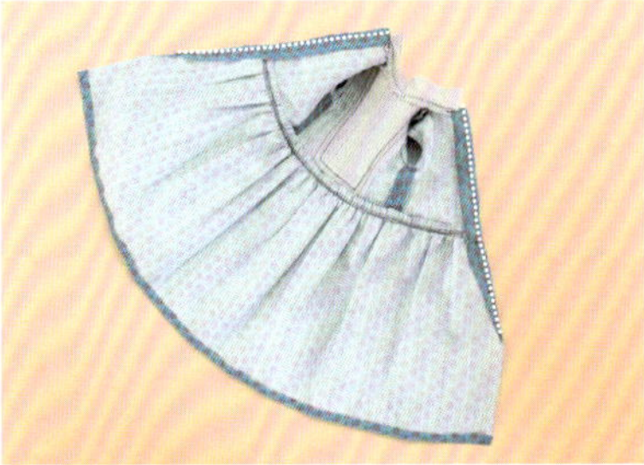

缝合后开衩。

42

从裙摆至“开衩止处”，正面向内缝合后中心。

43

熨烫摊开缝份，翻到正面。

44

穿 1 根绣花线，在领子边缘加入锁链绣。

45

后开襟侧缝接按扣，完成。

37. Sew the waist of the bodice and skirt. *38.* Iron the seam up to the bodice. *39.* Sew the edge of the bodice from the front.
40. Fold the seam down to where the openings should end and iron. *41.* Sew the back opening.
42. Pin the two sides together and sew. *43.* Unfold the seam and iron. Turn right side out.
44. Use one embroidery thread to chain-stitch on the collar for decoration. *45.* Fasten the snap at the back opening.

Blouse

罩衣

袖子的衣褶呈立体造型的罩衣。
无花边的简洁款式，穿着方便。

条纹薄棉布	S	20cm×17cm	8mm 花边	S 领子 12cm、袖子 6cm×2 条
	M	25cm×20cm		M 领子 16cm、袖子 8cm×2 条
	L	55cm×20cm		L 领子 24cm、袖子 12cm×2 条
纯色薄棉布	S	12cm×6cm	2.5mm 纽扣	S、M 2颗 L 6颗
	M	15cm×10cm	按扣	S、M 2组 L 5组
	L	20cm×15cm		

1

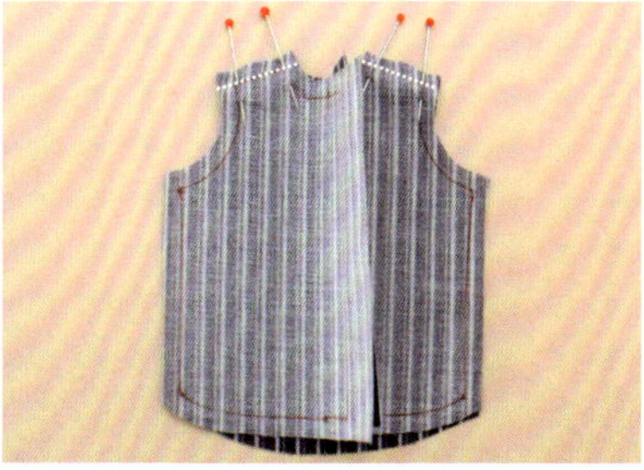

对照纸型裁剪各布件，布边涂抹锁边胶处理。正面对合前衣片和后衣片，缝合肩部。

2

熨烫摊开缝份。

3

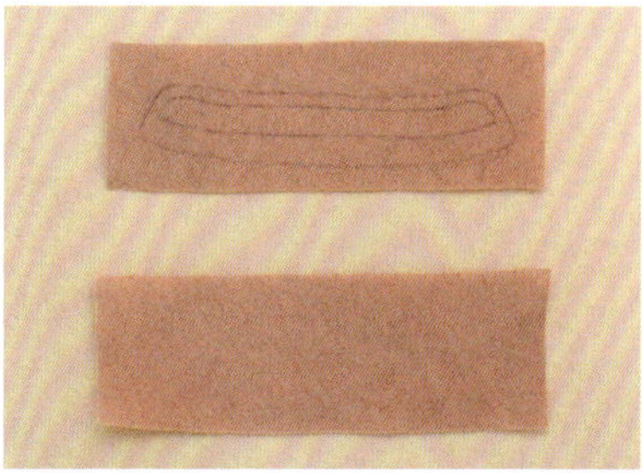

将领子的纸型描印于布料，稍大一圈裁剪，准备另一片同样尺寸的布料。

4

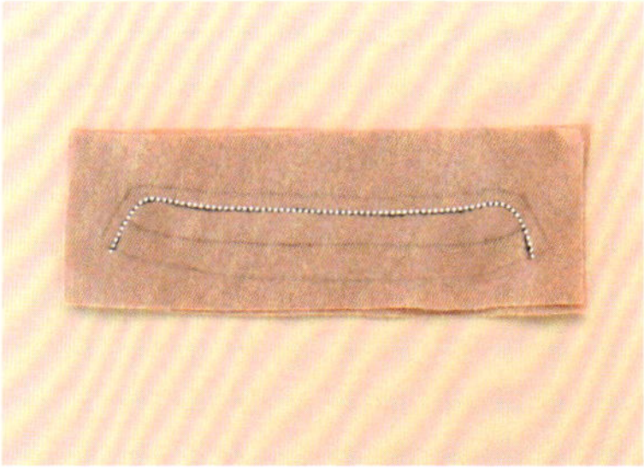

正面对合两片布料，缝合外侧的成品线。

5

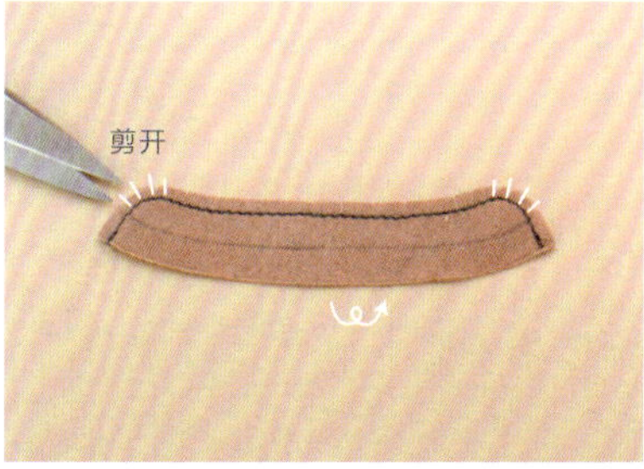
剪开

留下缝份之后裁剪，沿着弧线加入细密剪口。

6

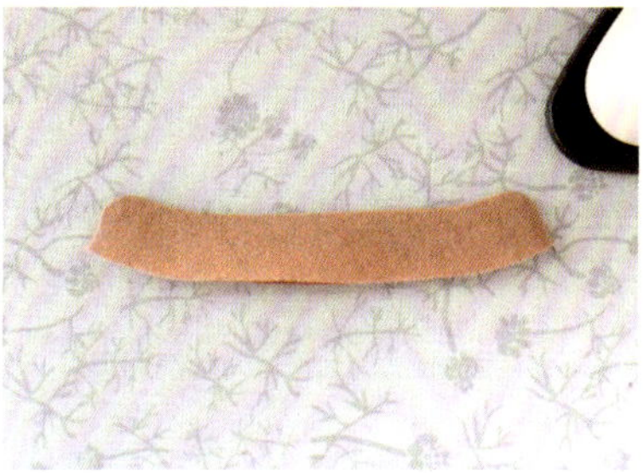

翻到正面，弧线整齐翻出，熨烫整齐。

7

剪开

衣片颈部周围的缝份加入细密剪口。

8

胶

缝份涂抹布料胶，放上领子。

9

正面向内折叠前衣片。

The shape of the sleeves looks neat and stylish thanks to the darts.
Widthout the lace, it can be a more basic item.

{ materials : cotton/stripe, cotton/plain, 8mm lace, 2.5mm buttons, snaps }

1. Cut out all the parts. Put fray-stopper glue on the edges. Match the front and back by the shoulders and sew.
2. Unfold the seam and iron. *3.* For the collar take two pieces of the same size and draw the collar on one piece.
4. Take the collar pieces and match the edges and sew the white lines as pictured.
5. Cut the shape of the collar and cut small cuts in the seam on the round. *6.* Turn the collar pieces inside out and iron. *7.* Make small cuts on the seam.
8. Put the fabric glue on the edges. *9.* Match the collar to the neck opening and fold the sides of the front over the edges of the collar.

10

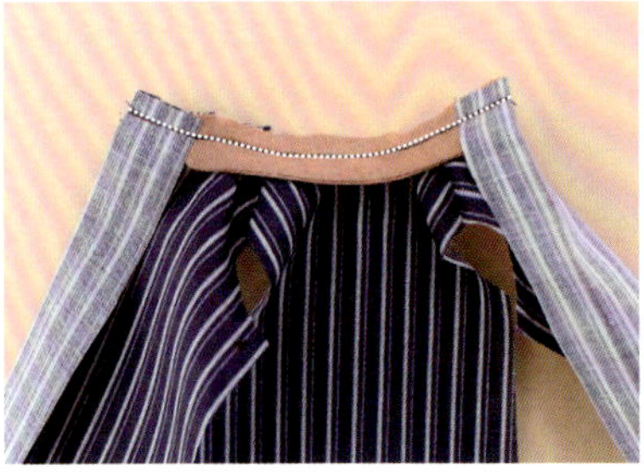

缝合颈部周围。

11

剪开

领子的缝份加入细密剪口。

12

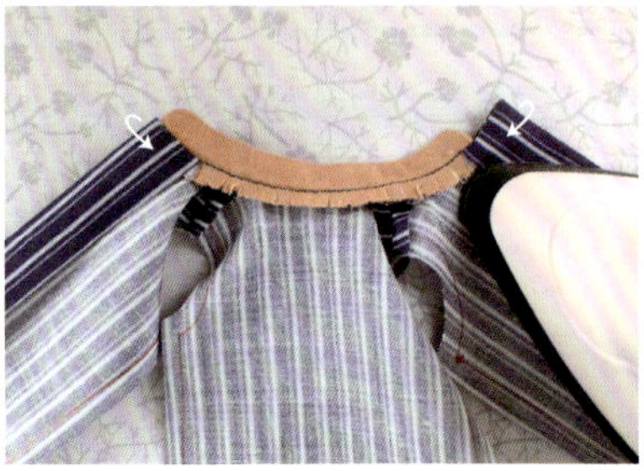

前衣片翻到正面，立起领子，熨烫整齐。

13

从已折入前衣片的前开襟的下摆开始至衣片的领子底边，明线车缝。

14

缝合前开襟、领子周围衣片、前开襟。

15

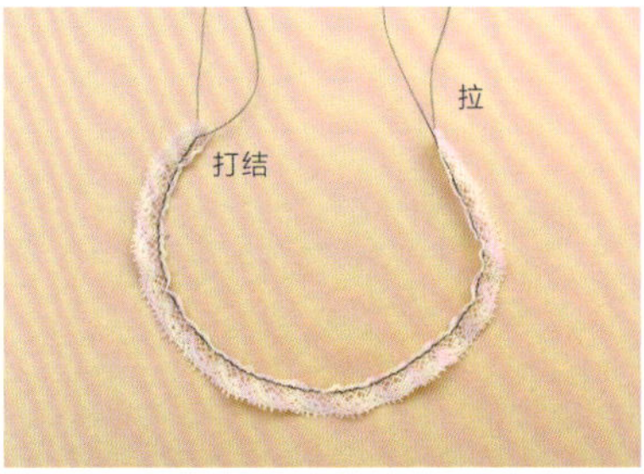
拉
打结

准备领子的花边。针脚调整为约 2.5mm 宽度，单线缩褶车缝花边的直线侧（未隆起部分）的边缘（参照第 90 页）。

16

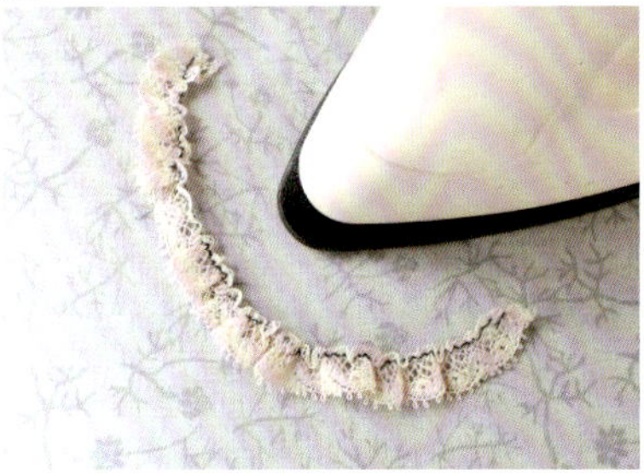

对齐领子，前中心的花边缝接位置的宽度进行缩褶，熨烫整齐。

17

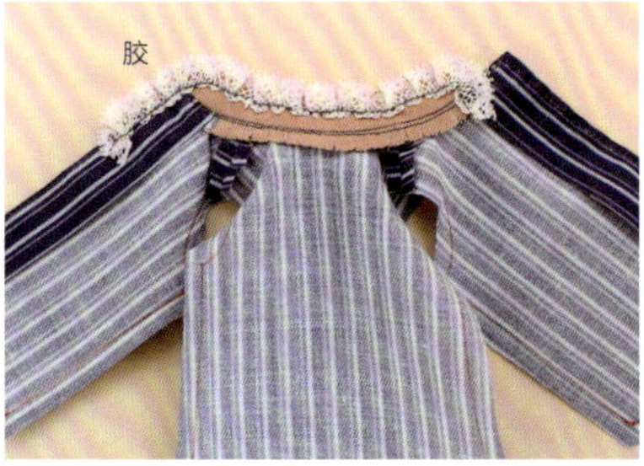
胶

用布料胶预固定花边。

18

缝接花边。

10. Sew the neck opening. *11.* Make small cuts on the seam. *12.* Turn right side out and iron as shown.
13-14. Sew the white line as shown. *15-16.* Gather the lace until the width fits the collar. [refer P.90]
17. Attach the lace with fabric glue. *18.* Sew the lace.

19

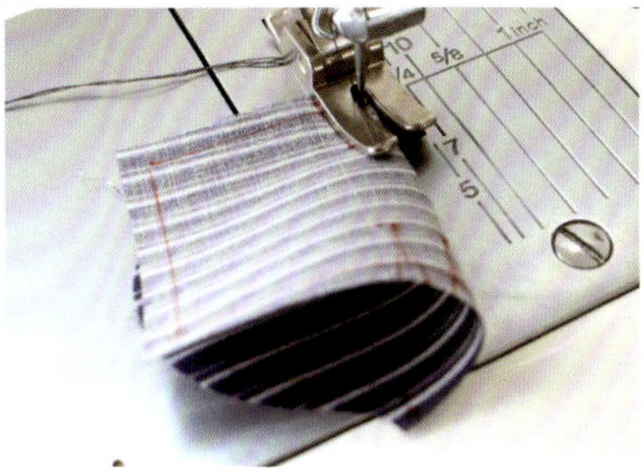

将袖子的衣褶位置逐个正面向内对折缝合，加入 4 个衣褶。

20

衣褶的缝份分别朝向内侧，熨烫压实。

21

※（仅 L 码从此步骤开始直接转至步骤 34 的袖口改造）正面向内缝合袖头和袖口。

22

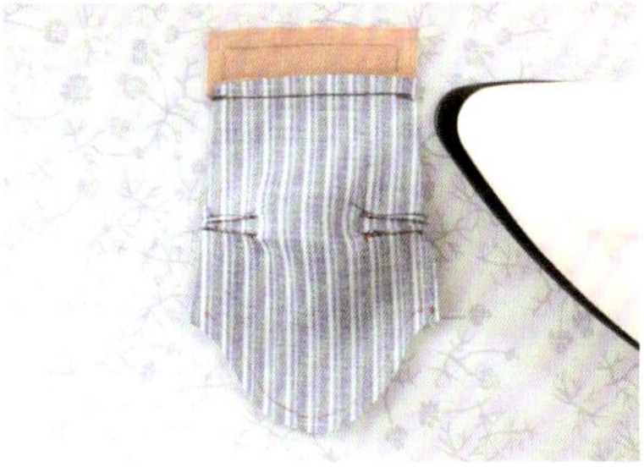

缝份压向袖头侧，翻折袖头。

23

熨烫三折边袖头，包住缝份。

24

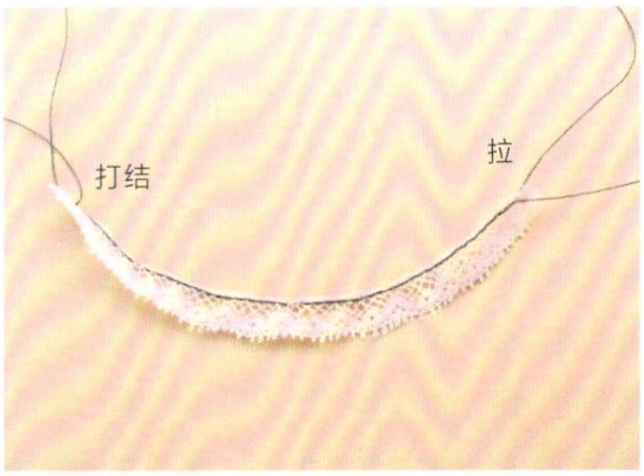
打结
拉

准备袖口的花边。针脚调整为约 2.5mm 宽度，单线缩褶车缝花边的直线侧的边缘。

25

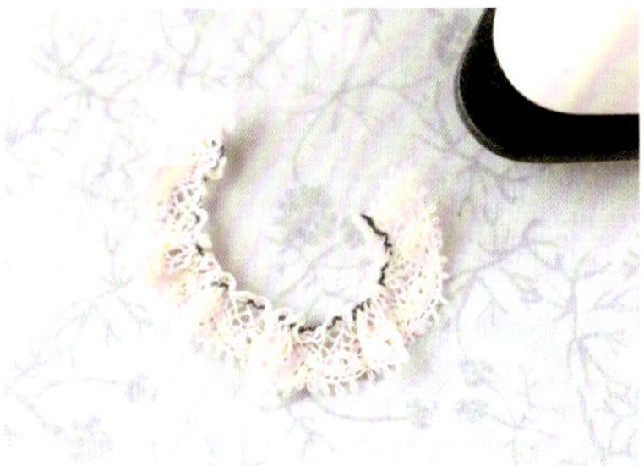

对齐袖头宽度进行缩褶，熨烫整齐。

26

胶

用布料胶预固定花边。

27

缝接花边，袖头整体明线车缝。

19. Sew the darts of sleeves. *20.* Iron the darts. *21.* Match the sleeve openings and the cuff and sew. [Please jump to image 34 in the case of <L> size]
22. Iron the seam of the cuff. *23.* Fold the cuff around the seam.
24-25. Gather the lace until the width fits the sleeve opening and iron. *26.* Attach the lace with fabric glue. *27.* Sew the cuff.

28

打结
拉

针脚调整为约 2.5mm 宽度，从袖山的缝份的拼合标记至拼合标记单线缩褶车缝。对齐衣片的袖窿宽度，进行缩褶（参照第 90 页）。

29

正面向内缝合衣片的袖窿和袖山。缝份压向袖子侧，熨烫整齐。

30

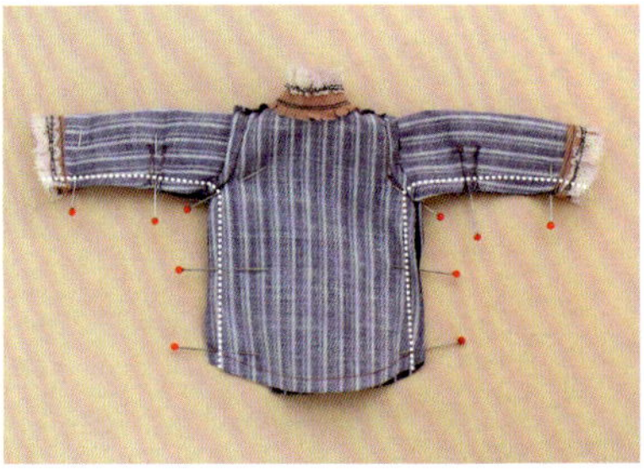

正面对合前衣片和后衣片，缝合袖口、侧身、下摆。

31

剪开

侧身的缝份加入剪口。翻到正面，熨烫摊开缝份。

32

将下摆的缝份熨烫折向内侧，缝合。

33

前开襟缝接按扣，袖头缝接装饰纽扣，S 码和 M 码完成。

34

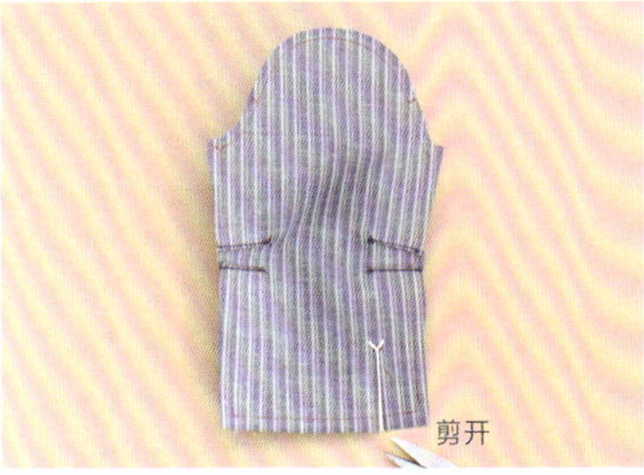
剪开

L 码在袖口制作开衩。对照纸型，左右袖口加入剪口。

35

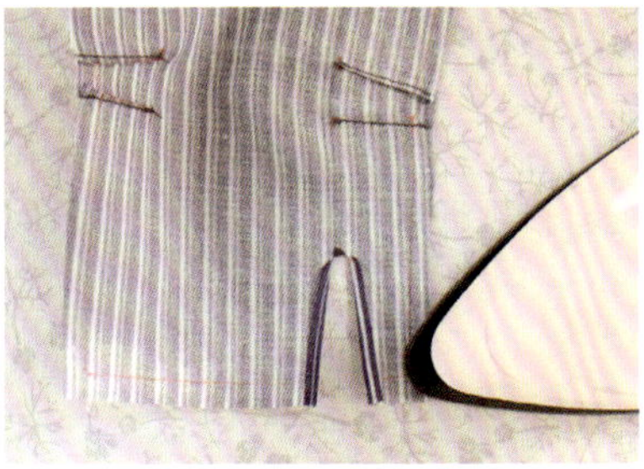

按照 2mm 宽度将剪口折入内侧，熨烫整齐。

36

明线车缝。

28. Gather the shoulders until the width fits the armhole. [refer P.90]
29. Match the side edge of the sleeve to the bodice and gradually sew the shoulder of the sleeve to the armhole, matching as you go.
30. Pin the sides of the bodice and sleeves and sew. *31.* Make cuts in seam allowance of the pits. Turn right side out and unfold the seam.
32. Iron the hem and sew. *33.* Fasten snap at the opening and attach the button to the cuff.
34. For <L> size. Cut the sleeve opening as shown. *35.* Fold the seam down and iron. *36.* Sew the sleeve opening.

37

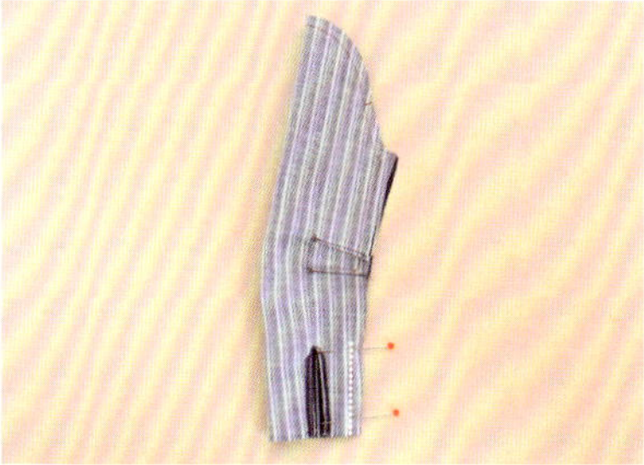

正面向内折入袖子，缝合至距离袖口2cm 左右位置。

38

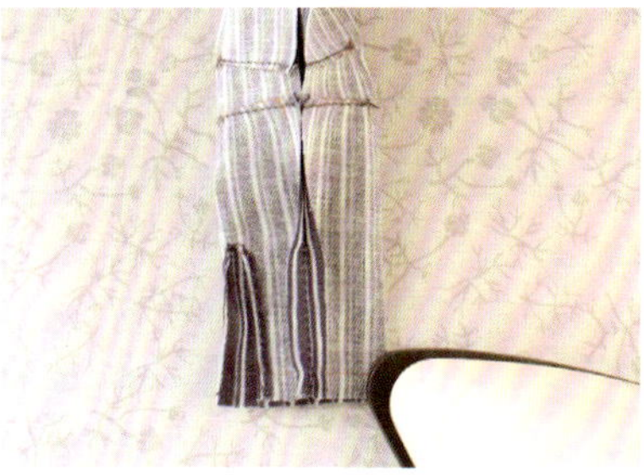

熨烫摊开缝份。

39

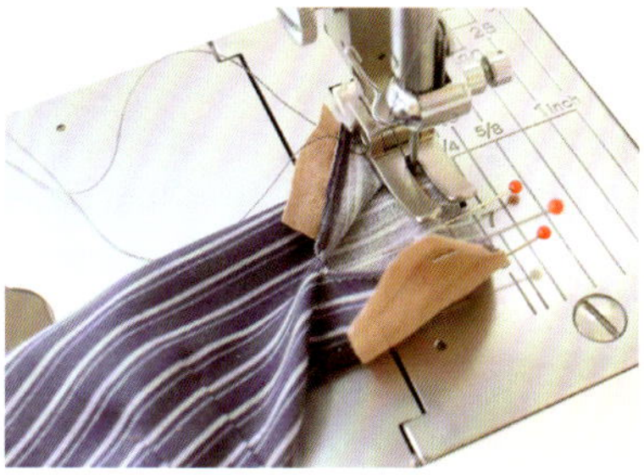

袖子翻到正面，袖头和袖口正面向内缝合。

40

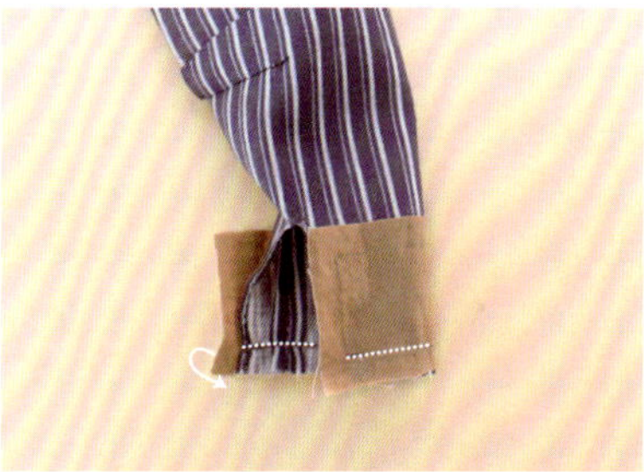

袖头翻到正面。

41

剪掉

正面向内缝合另一片袖头和之前的袖头，剪掉缝份的边角。

42

袖头翻到正面，熨烫整齐。

43

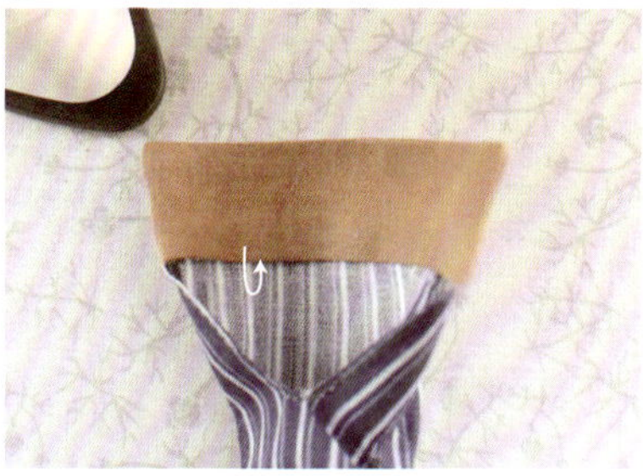

缝份熨烫折入内侧，熨烫整齐。

44

胶

袖口的花边缩褶，预固定于袖头之后缝合，袖头整体明线车缝。

45

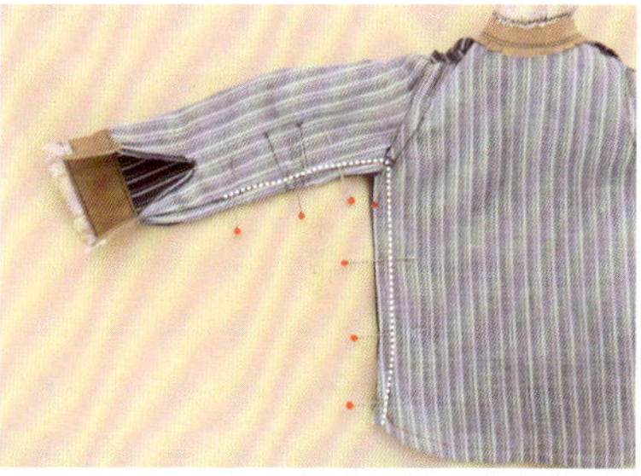

袖子翻到里衬，正面对合前衣片和后衣片，缝合袖口、侧身、下摆。最后，袖头缝接按扣和装饰纽扣。

37. Sew 2cm from the edge as shown. *38.* Unfold the seam and iron. *39.* Turn the right side out. Match the sleeve to the cuff and sew. *40.* Match the other cuff and sew. Turn right side out and iron. *41.* Cut the corners as shown for easier folding. *42-43.* Fold the seam inside and iron. *44.* Gather the lace until the width fits the sleeve opening and sew. *45.* Turn the sleeve inside out and sew as shown. From here, please go back to image 31.

Skirt

短裙

大小适中的抽褶短裙初学者正好练手。
本文采用红色布料制作，但如果制作本白色、黑色系等颜色同样漂亮。

棉亚麻布	S 38cm×15cm M 40cm×20cm L 90cm×30cm	4mm 纽扣	S、M 2 颗 L 4 颗
		按扣	2 组

1

对照纸型裁剪各布件，布边涂抹锁边胶处理。裙片正面向内对合，缝合侧身。

2

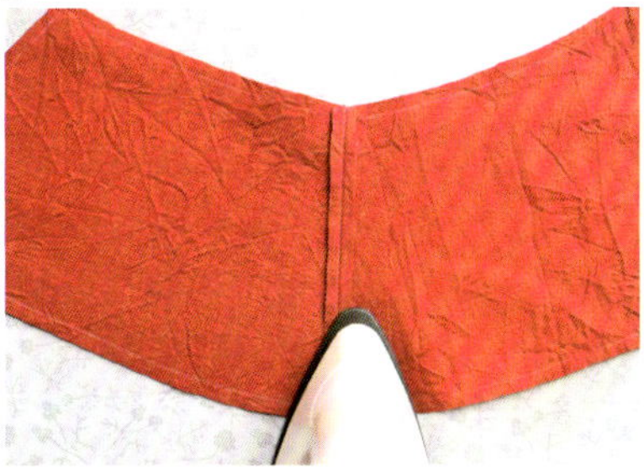

熨烫摊开缝份。

3

正面向对缝合前育克和后育克。制作两片，表布及里衬。

4

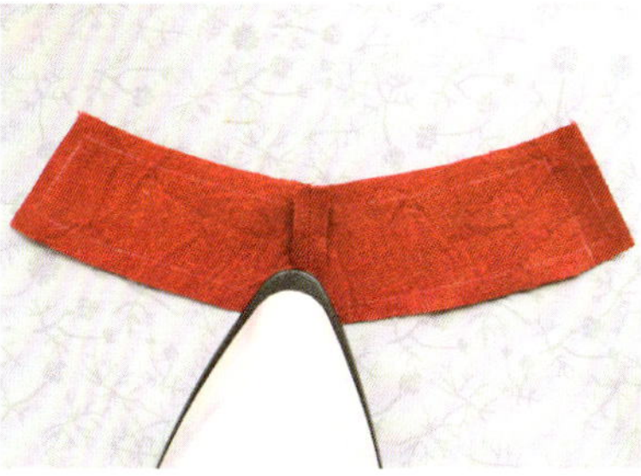

熨烫摊开缝份。

5

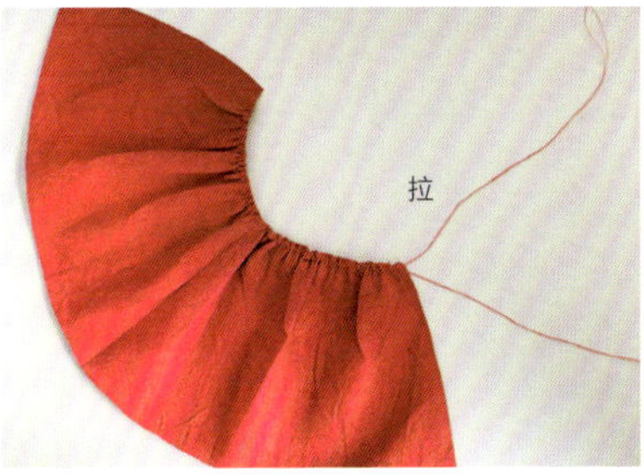
拉

针脚调整为 3mm 宽度，双线缩褶车缝裙片腰围的缝份（参照第 90 页）。

6

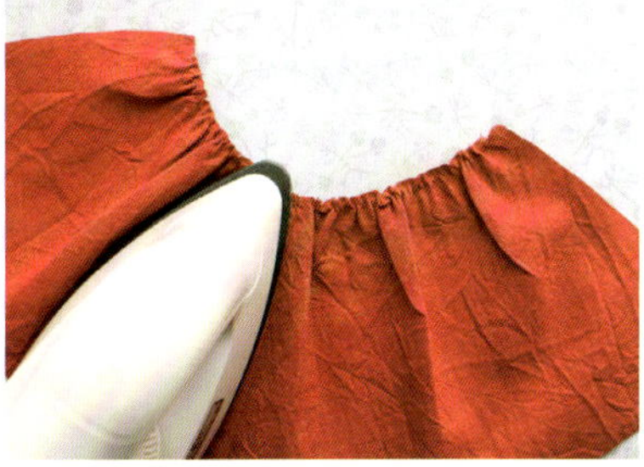

对齐育克的宽度进行缩褶，熨烫整齐。

7

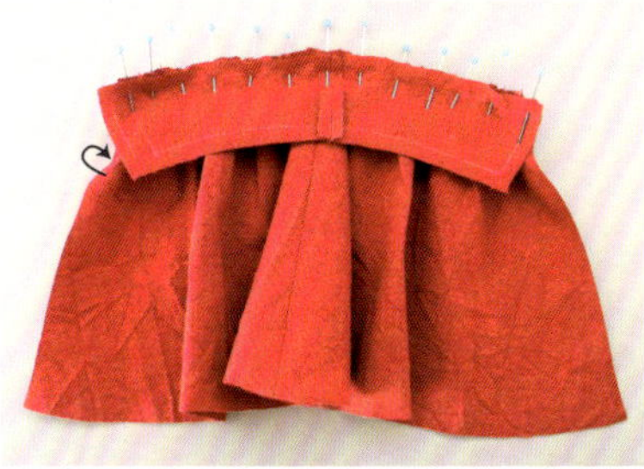

正面对合裙片和育克，将裙片一侧（后育克侧）的缝份折向内侧。

8

正面向对缝合裙片和育克。

9

缝份压向上方，熨烫压实。

Modest gathering is recommended for beginners.
This skirt will look wonderful in any color.

{ materials : cotton linen, 4mm buttons, snaps }

1. Cut out the parts. Put fray-stopper glue on the edges. Sew the skirt together. *2.* Unfold the seam and iron.
3. Sew together front of the waist yoke. Make 2 sets. *4.* Unfold the seam and iron.
5. Gather the waist of the skirt, until the width fits the waist yoke. [refer to P.90] *6.* Iron the gathering.
7. Match the one yoke piece to the skirt. Fold the side seam over the yoke in one side. *8.* Sew the waist. *9.* Iron the seam up.

10

育克上方正面对合另一片育克，并缝合。

11

剪开

剪掉缝份的边角，加入剪口。育克翻入正面。

12

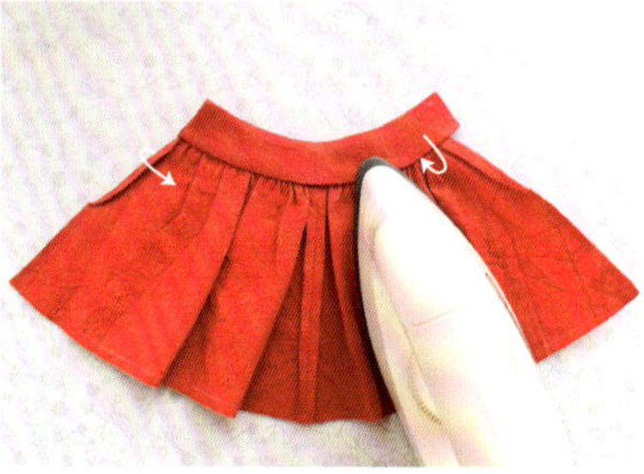

育克的缝份压向内侧，熨烫整齐。倾斜折入裙片的开衩（“开衩止处”拼合标记的正下方为止）。

13

育克明线车缝，缝合后育克的开衩。

14

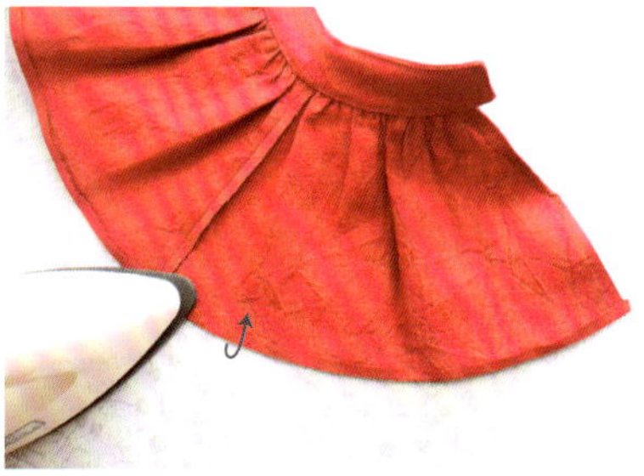

裙摆的缝份折向内侧，熨烫压实。

15

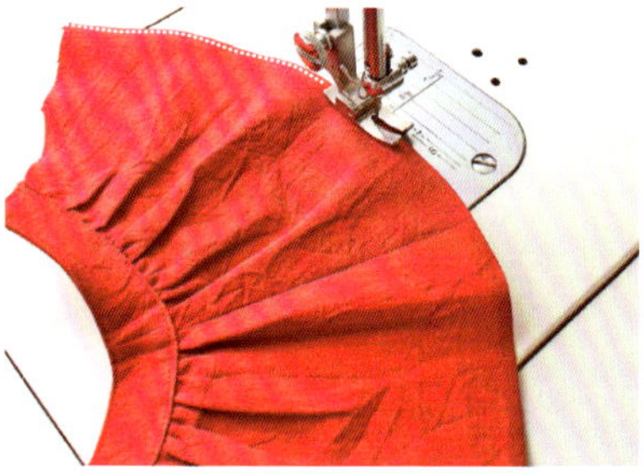

缝合裙摆。

16

从裙摆至“开衩止处”，正面向内缝合侧身。

17

熨烫摊开缝份。

18

水洗后自然干燥，缝接装饰纽扣即完成。

10. Match the other yoke piece to the attached yoke and sew as shown. *11.* Cut the corners and along the seam. Turn the yoke right side out. *12.* Fold the seam inside and iron. *13.* Sew the yoke and opening. *14.* Fold the hem of the skirt. *15.* Sew the hem. *16.* Sew the opening. *17.* Unfold the seam and iron. *18.* Wash the skirt in water and dry. This will give the cloth a natural finish. Fasten snaps and attach the buttons.

Vest & Corsage

背心 & 胸花

使用棉或亚麻布料制作，春夏秋均适合穿着的背心。
胸花是精心点缀，这样背心不会太单调。

细灯芯绒布	S 16cm×12cm M 22cm×15cm L 30cm×20cm	薄棉布	S 16cm×10cm M 22cm×10cm L 30cm×18cm
2.5mm 纽扣	S、M 5颗 L 6颗	弹簧扣（凸）	S、M 2颗 L 3颗
缠纸铁丝		徽章胸针	

1

正面对合表布的前衣片和后衣片，缝合肩部。

2

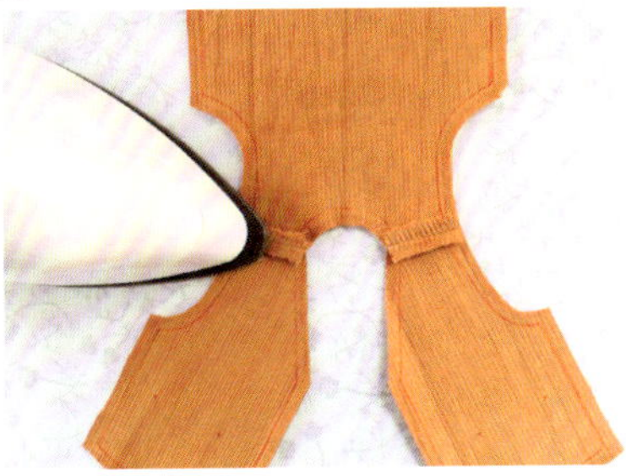

熨烫摊开缝份。

3

里衬同样，正面对合前衣片和后衣片，缝合肩部。

4

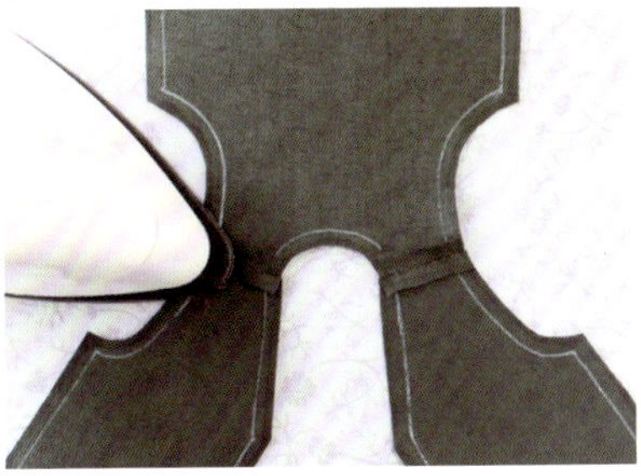

熨烫摊开缝份。

5

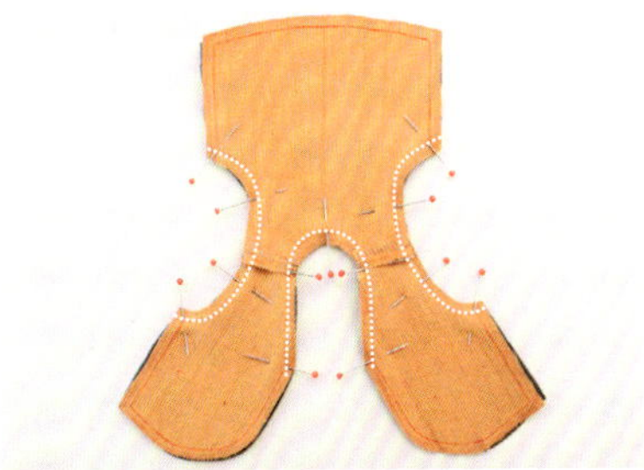

正面对合表布和里衬，缝合颈部周围和袖口。

6

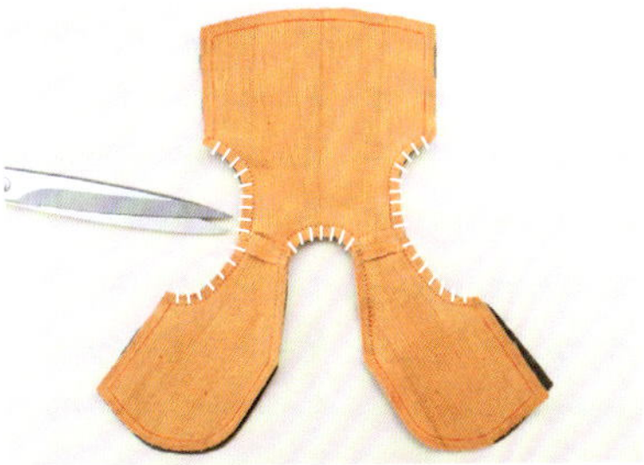

缝份加入细密剪口。

7

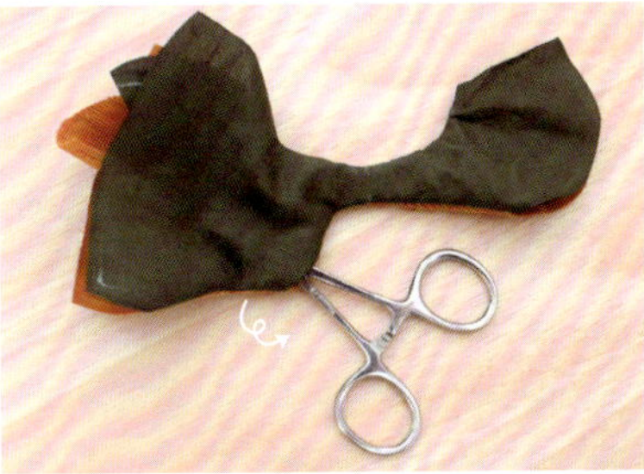

使用钳子等翻到正面。

8

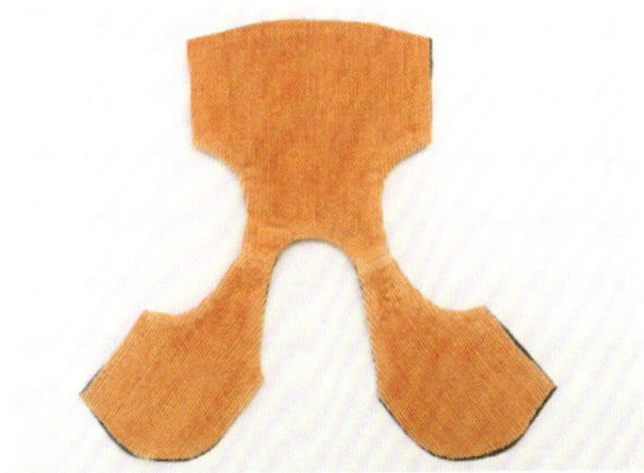

用锥子等工具将边角及弧线调整齐，熨烫整齐。

9

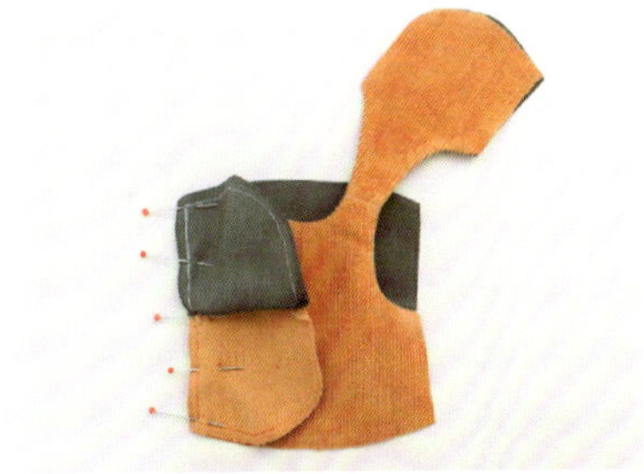

正面对合前衣片和后衣片的各表布侧身，并用珠针固定。同样，正面对合前衣片和后衣片的各里衬侧身，并用珠针固定。

To match the season, choose linen or cotton for a light feel.
The corsage adds a touch of style to the outfit.

{ materials : corduroy, cotton, 2.5mm buttons, hooks }

1. Match the front and back by the shoulders and sew. *2.* Unfold the seam and iron.
3. Match the front and back of the lining by the shoulders and sew. *4.* Unfold the seam and iron.
5. Match the outside and the lining and sew as shown. *6.* Cut along the seam. *7.* Turn right side out. *8.* Try to make the edges look neat.
9. Match the sides of the outer side and pin together up to armhole. Fold the lining up so that back and front meet above the armhole and pin.

10

分别缝合各表布的侧身、各里衬的侧身。

11

熨烫摊开表布及里衬的侧身的缝份。正面对合表布及里衬，从前开衩开始至下摆用珠针固定。

12

保留下摆的翻口，缝合左右前开襟。

13

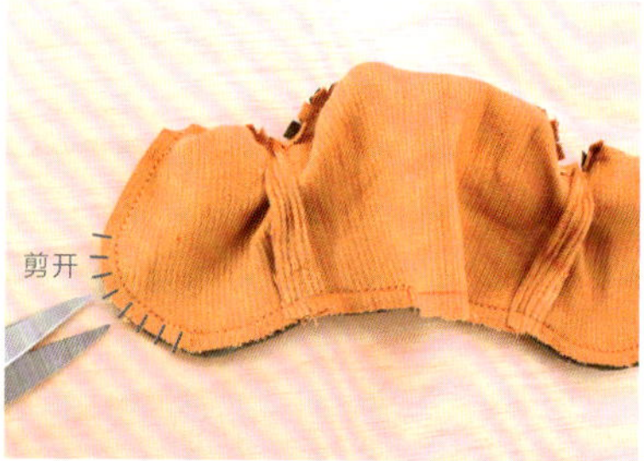
剪开

剪掉缝份的边角，沿着弧线加入剪口。

14

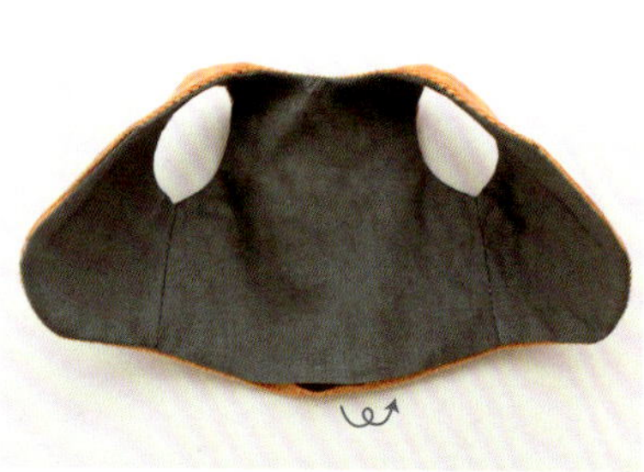

用钳子等工具，从翻口逐渐翻到正面。用锥子等工具将边角及弧线调整齐，熨烫整齐。

15

缭缝固定翻口。

16

折入袋口的缝份，熨烫压实。

17

缝合口袋口。

18

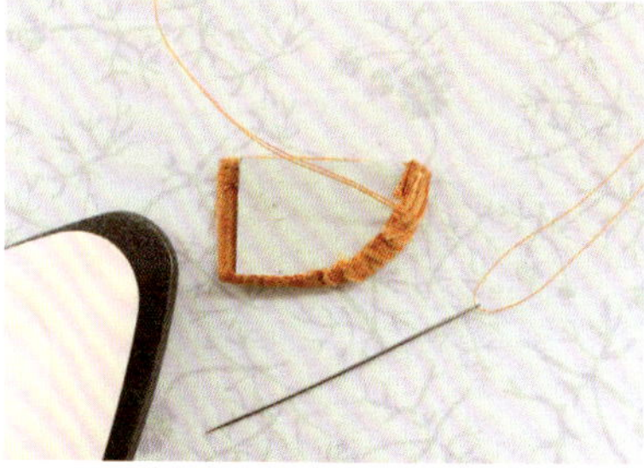

在口袋弧线的缝份，进行 3mm 宽度的拱缝。对照纸型拉线制作弧线部分，熨烫压实缝份。

10. Repeat on other side and sew both sides. *11.* Match the lining and front side by the edges and pin. *12.* Sew as shown. *13.* Cut along the seam. *14.* Turn right side out and iron. *15.* Sew with blind stitch. *16-17.* Fold the top of the pockets, iron and sew. *18.* Use running stitch to sew the bottom of the pocket and fold.

19

另一侧的缝份同样折入，熨烫整齐。

20

口袋的缝份涂上布料胶，预固定于衣片，并缝合。

21

前开襟缝接弹簧扣，另一侧制作线袢（参照第 91 页）。

22

缝接装饰纽扣，完成。

23

使用表布的多余布料，制作胸花。剪出 3 片边长 1cm 的正方形。准备 3cm 人造花中使用的缠纸铁丝。

24

剪掉正方形布料的边角，边缘涂抹锁边胶处理。

25

3 片重合，用线止缝正中央。

26

对折缠纸铁丝，止缝于里衬。

27

将布料揉皱，将徽章胸针或别针止缝于背面。

19. Fold the side seam. *20.* Attach the pockets on the bodice with glue and sew. *21.* Attach the hooks and make thread loops. [refer to P.91]
22. Attach the front buttons. *23.* Cut three 1cm squares. *24.* Cut the corners and put fray-stopped glue on the edges.
25. Sew the three pieces together in the center. *26.* Attach the wire. *27.* Crumple the pieces. Attach the pin.

Trousers

裤

与背心成套的裤子。
建议使用棉布或亚麻布洗旧后制作。

细灯芯绒布	S 20cm×18cm M 30cm×25cm L 40cm×35cm	袋兜用薄棉布	S 6cm×3cm M 8cm×5cm L 10cm×7cm
按扣	S、M 1组 L 2组		

1

裁剪各布件，布边涂抹锁边胶处理。袋兜布正面对合放于前裤片的袋口，缝合袋口。

2

剪开

缝份加入剪口。

3

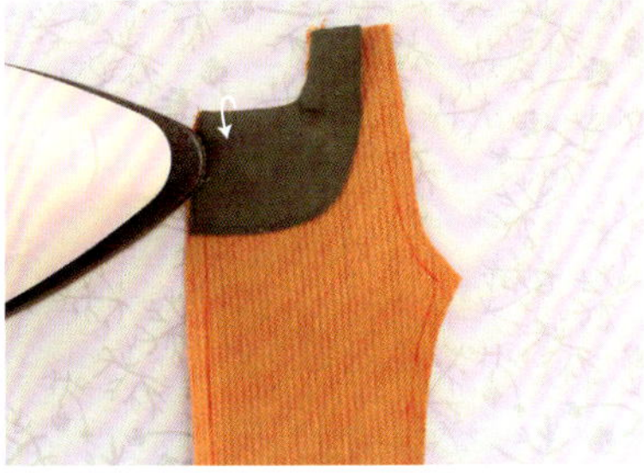

袋兜布翻到反面，熨烫整齐。

4

袋口明线车缝。

5

口袋的侧边布正面向内重合于袋兜布，缝合。

6

此时，注意避免将底部的裤片一起缝合。

7

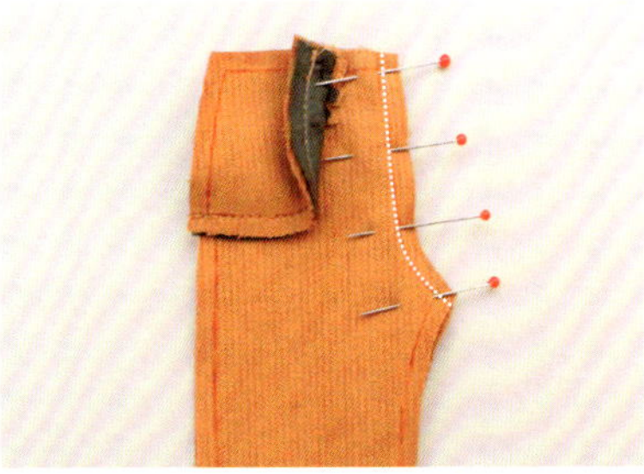

正面对合左右前裤片，缝合上裆。

8

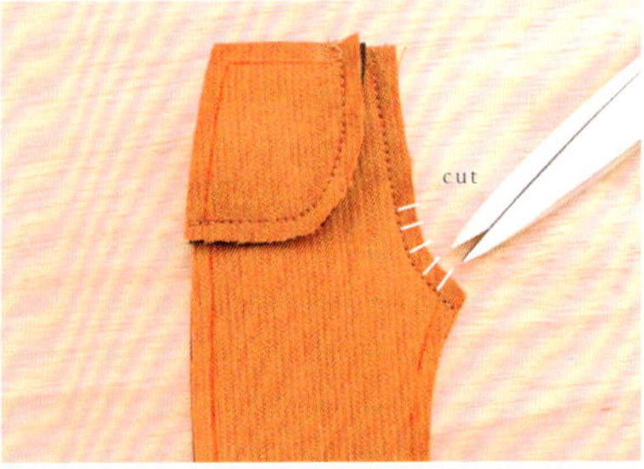
cut

沿着缝份的弧线加入剪口。

9

熨烫摊开缝份。

Match these trousers to the vest for a full look.
Wash them in water for a more realistic look for cotton or linen.

{ materials : corduroy, cotton, snaps }

1. Cut out all parts. Put fray-stopper glue on the edges. Match the pocket and lining the outside of the front pieces and sew.
2. Cut along the seam. *3-4.* Turn the pocket inside, iron and sew. *5.* Match the back of the pocket and lining and sew.
6. Sew together the pocket back and lining. Make sure not to sew it to the front of the pants.
7. Sew together the front pieces in the center. *8.* Cut along the seam. *9.* Unfold the seam and iron.

10

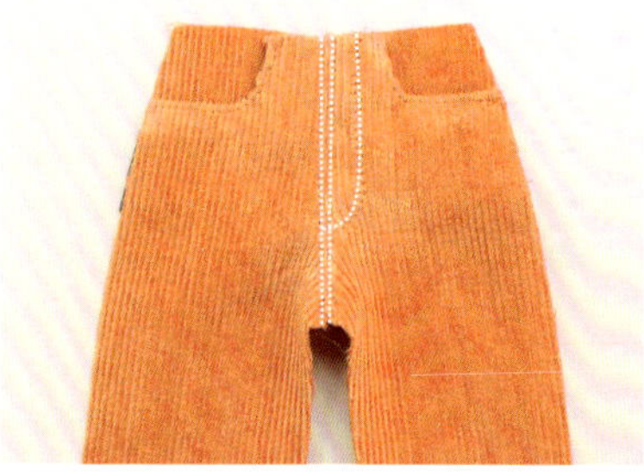

上裆的左右侧明线车缝，并在左侧前方明线车缝。

11

正面对合前裤片和后裤片，缝合侧身。

12

熨烫摊开缝份。

13

侧身的左右侧明线车缝。

14

将裤脚的缝份熨烫折向内侧。

15

缝合裤脚。

16

正面对合裤片和腰带，缝合腰围。缝份压向上方。

17

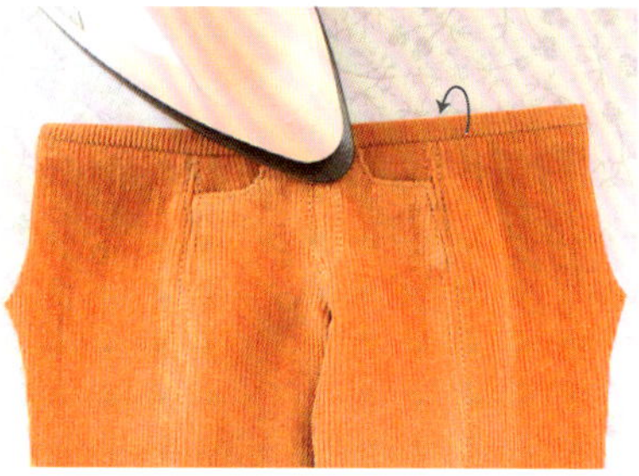

熨烫三折边腰带，包住缝份。布料较厚时，也可对折。

18

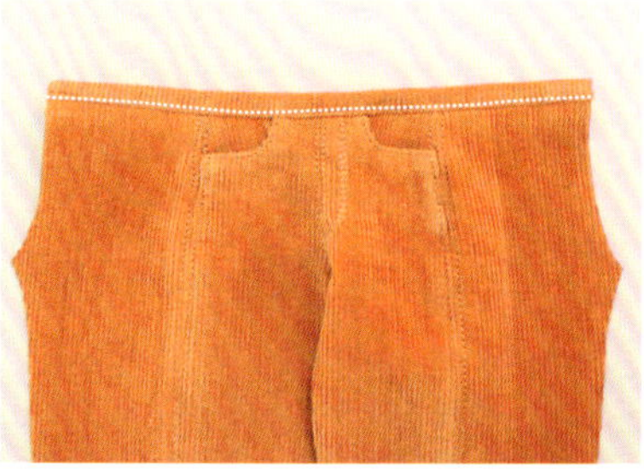

腰带侧明线车缝。

10. Stitch the front center as shown. *11.* Match the front and back of the pants and sew. *12.* Unfold the seam and iron.
13. Sew the sides on the outside. *14-15.* Fold the hem , iron and sew.
16. Match the pants and belt and sew. Iron the seam up. *17-18.* Fold the belt and sew.

19

cut

从反面看，在右侧后上裆的缝份“开衩止处”拼合标记位置加入剪口。

20

将剪口上方的后开衩的缝份折向内侧，熨烫压实。

21

缝合后开衩。

22

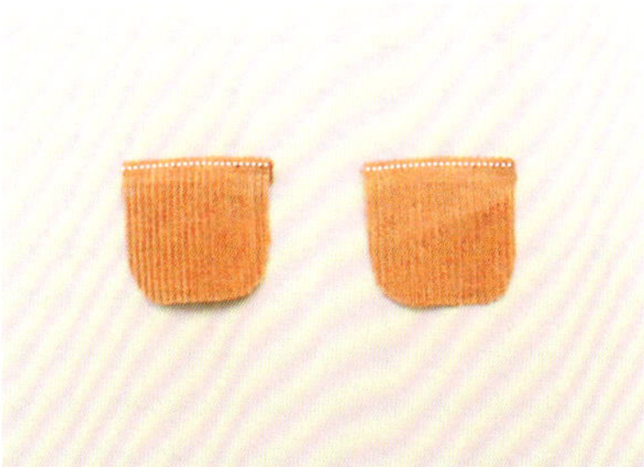

折入后口袋的袋口的缝份，并缝合。

23

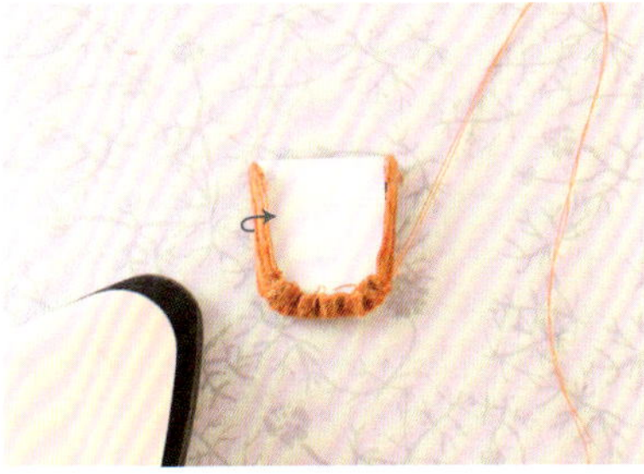

口袋弧线的缝份侧，进行 3mm 宽度的拱缝。对照纸型拉线制作弧线部分，熨烫压实缝份。

24

口袋的缝份涂抹布料胶，预固定于后裤片，并缝接。

25

正面对合后裤片，缝合后上裆（至“开衩止处”）。熨烫摊开缝份。

26

缝合下裆。

27

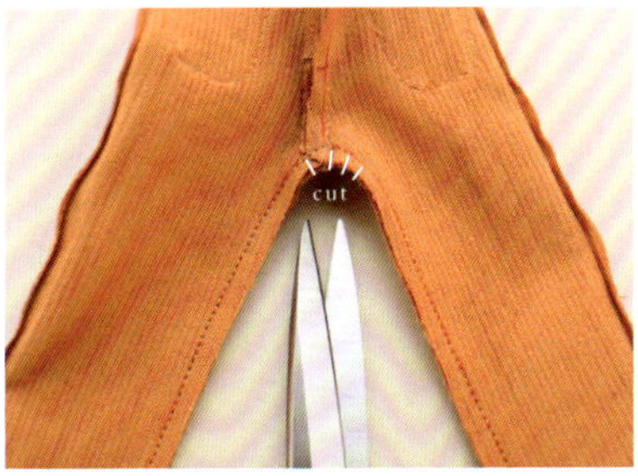
cut

下裆的缝份加入剪口，翻到正面。后开衩缝接按扣，即完成。

19. Cut the side of the back opening as shown. *20-21.* Fold the seam and sew. *22.* Fold the top of the pocket.
23. Sew the bottom of the pocket with running stitch and iron the seam. *24.* Attach the pocket and sew.
25. Sew the back opening and unfold the seam, iron. *26.* Sew the inseam. *27.* Cut the seam and turn the pants right side out. Fasten a snap.

Coat
外套

飘逸甜美的外套。
使用棉布或亚麻布等搭配出季节感。

天鹅绒	S 30cm×20cm M 45cm×25cm L 75cm×50cm	里衬用薄纱布	S 18cm×4cm M 25cm×5cm L 35cm×7cm
弹簧扣（凸）	S、M 3颗 L 5颗	4mm 纽扣	S、M 1颗 L 3颗
按扣	L 2颗		

1

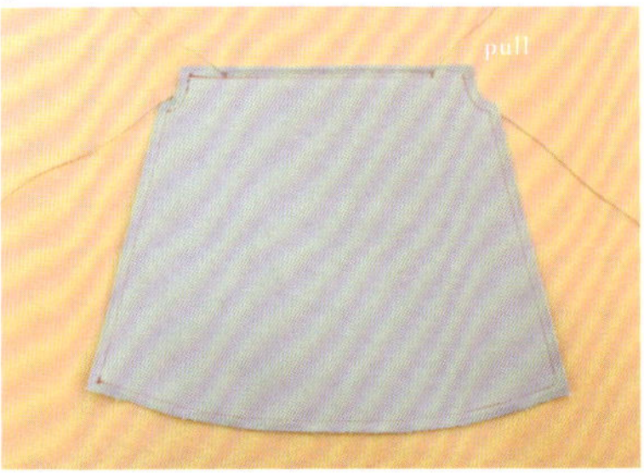
pull

各布边涂抹锁边胶处理。针脚调整为约3mm 宽度，从后衣片的过肩侧的缝份的拼合标记至拼合标记双线缩褶车缝。

2

对齐后过肩的宽度进行缩褶，熨烫整齐（参照第 90 页）。

3

正面向内缝合后衣片和后过肩。

4

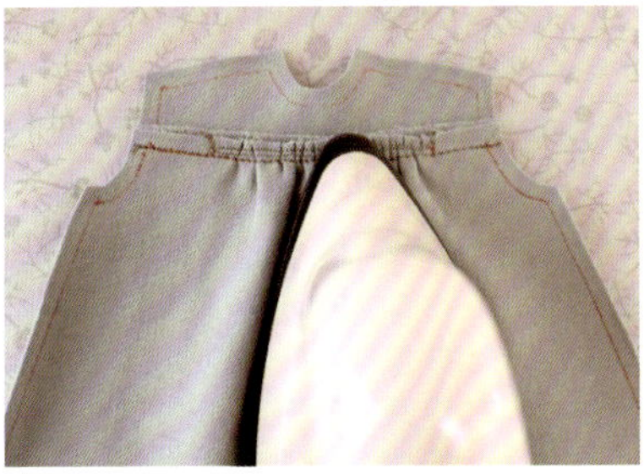

缝份压向过肩侧，熨烫整齐。

5

过肩侧明线车缝。

6

pull

针脚调整为约 3mm 宽度，从前衣片过肩侧的缝份的拼合标记至拼合标记双线缩褶车缝，对齐前过肩成品线的宽度进行缩褶。

7

正面向内缝合前衣片和前过肩。

8

缝份压向过肩侧，熨烫整齐。

9

过肩侧明线车缝。

This coat is a bit oversized for a more feminine look.
Use linen for a feel of spring or summer.

{ materials : velvet, cotton, hooks, 4mm buttons, for <L> snaps }

1-2. Cut out all the parts. Put fray-stopper glue on the edges. Gather the back until the width fits the back yoke. [refer to P.90]
3. Match the back yoke to the back and sew. *4-5.* Iron the seam up and sew.
6. Gather the front until the width fits the front yoke.[refer to P.92] *7.* Match the front yoke and front and sew. *8-9.* Iron seam up and sew.

10

正面对合前衣片和后衣片，缝合肩部。

11

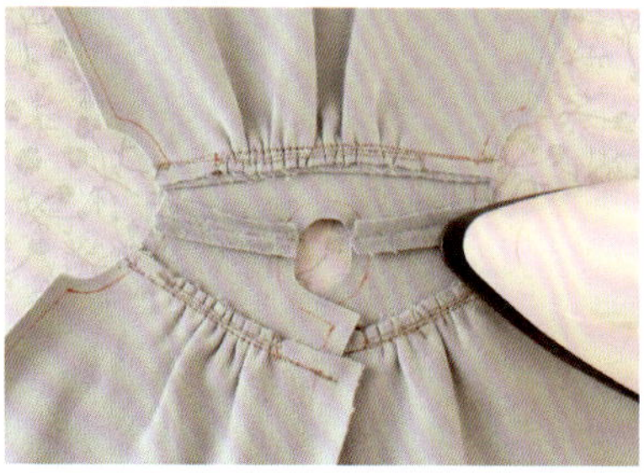

熨烫摊开缝份。

12

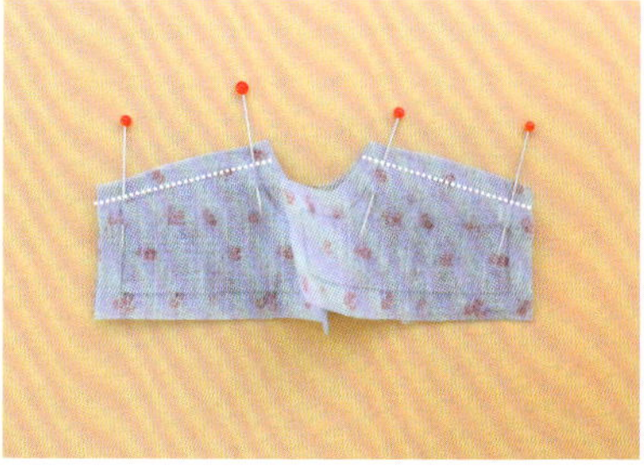

正面对合里衬的前衣片和后衣片，缝合肩部。

13

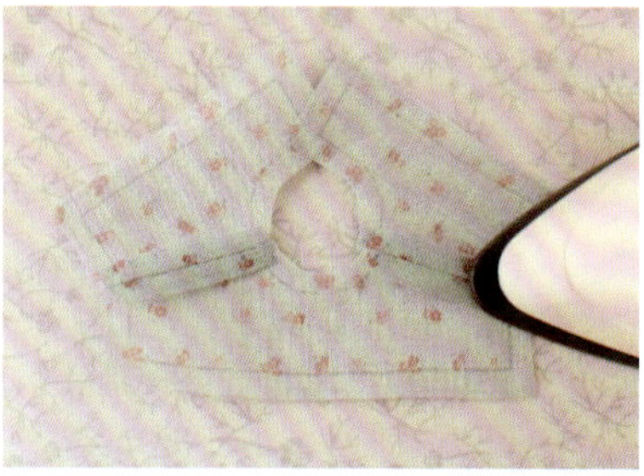

熨烫摊开缝份。

14

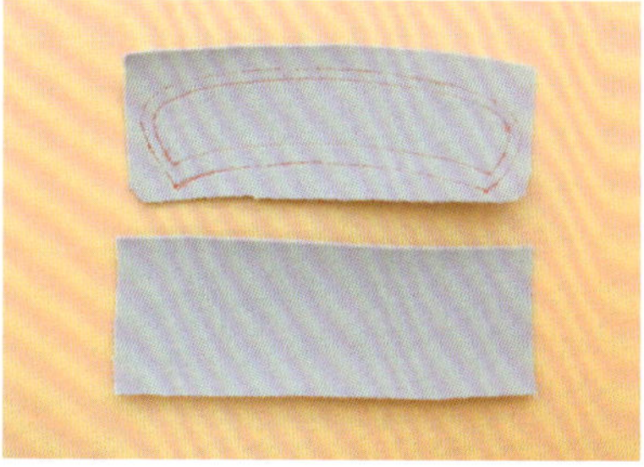

将领子描印于布料，按稍大一圈裁剪，并准备另一片相同尺寸的布料。

15

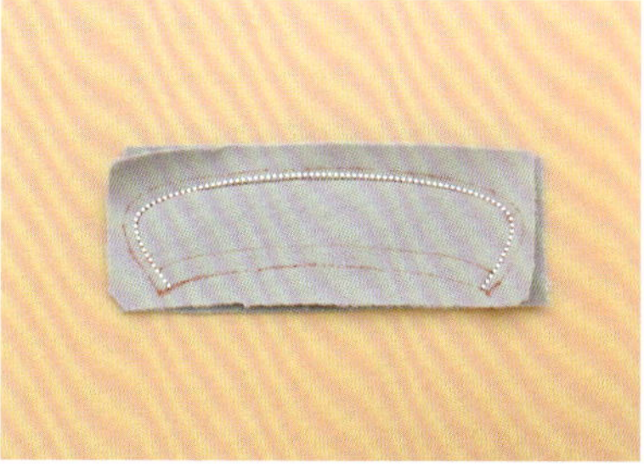

正面对合两片布料，缝合外侧的成品线。

16

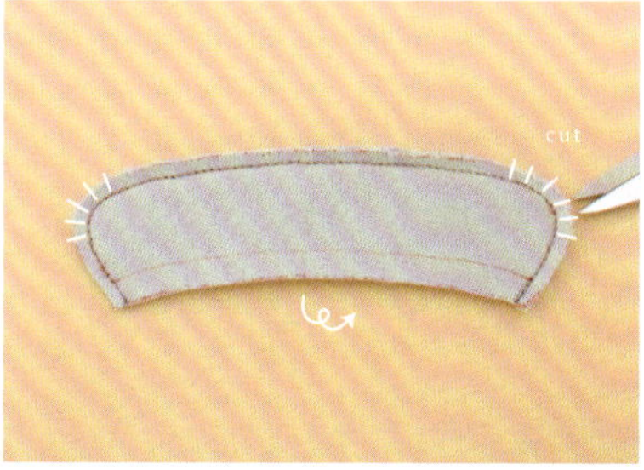
cut

留下缝份之后裁剪领子，沿着弧线加入剪口。

17

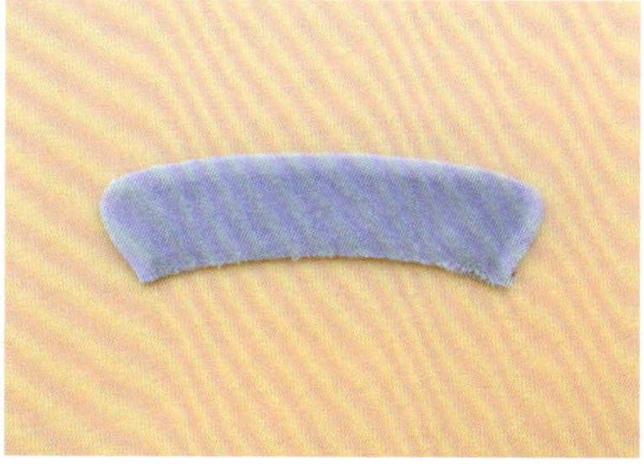

翻到正面，用锥子等将边角及弧线调整齐，熨烫整齐。

18

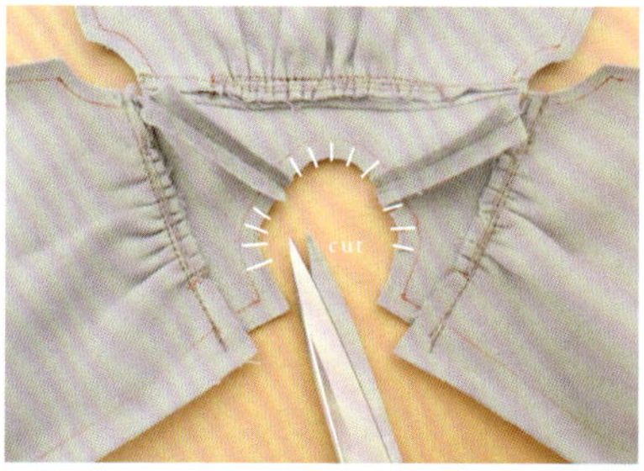
cut

过肩的颈部周围的缝份加入细密剪口。

10. Match the front and back of the yoke by the shoulders and sew.
11. Unfold the seam and iron. *12.* Match the front and back of the yoke by the shoulders and sew.
13. Unfold the seam and iron. *14.* For the collar take two pieces of the same size, and draw the collar on one piece.
15. Take the collar pieces and match the edges, sew. *16.* Cut the shapes of the collar and cut small cuts in the seam on the round.
17. Turn the collar piece right side out and iron. *18.* Make small cuts along the seam.

19

glue

缝份涂抹布料胶，将领子正面对合之后放上预固定。

20

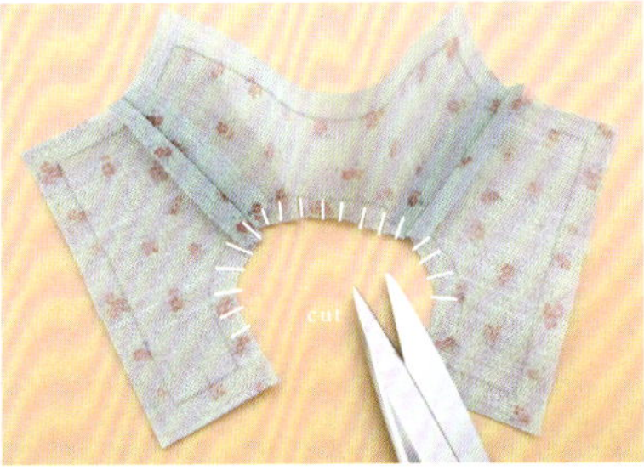
cut

里衬颈部周围的缝份加入细密剪口。

21

正面对合过肩和里衬，用布料胶预固定缝份，车缝缝合前开襟至颈部周围。领子的缝份加入剪口，翻到正面。

22

熨烫折入衣片的前贴边。里衬袖窿的缝份压向内侧，熨烫压实。

23

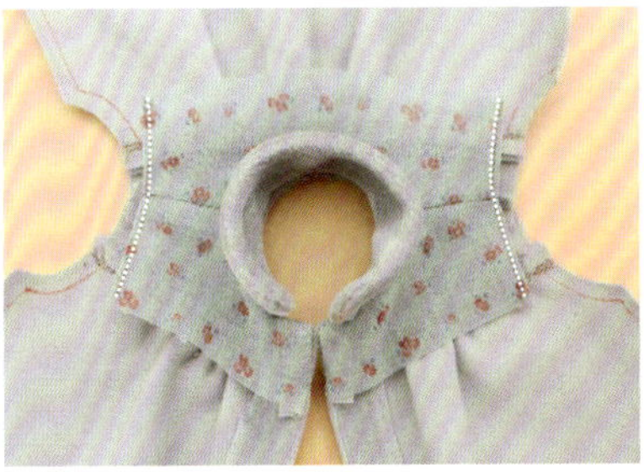

缝合里衬的袖窿（注意避免与底部的过肩一起缝合）。

24

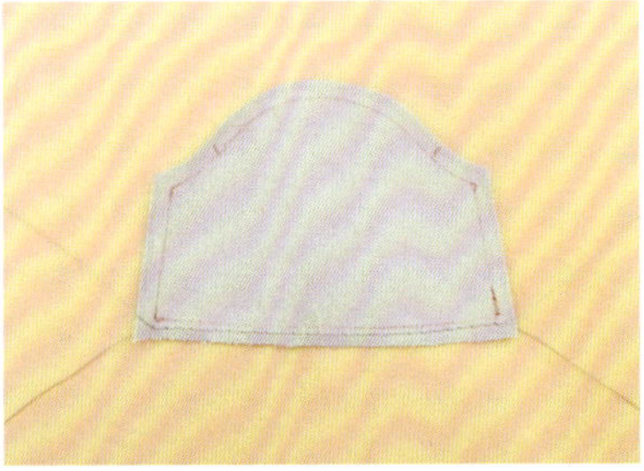

仅 L 码的袖口加入剪口。与罩衣的步骤 34~38 相同处理。接着，袖口缩褶。

25

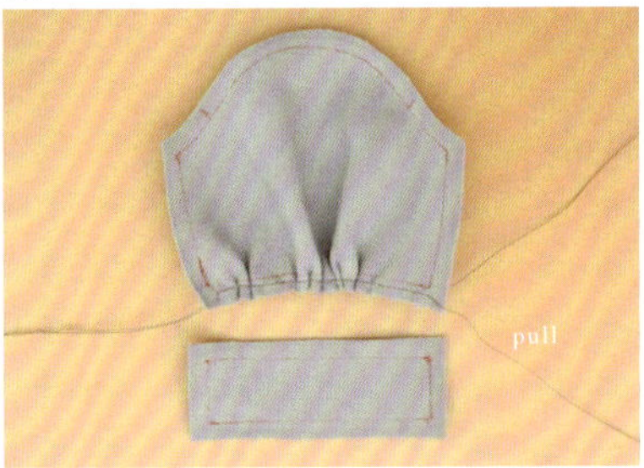
pull

针脚调整为约 3mm 宽度，单线缩褶车缝袖口，与袖头的宽度对齐进行缩褶。

26

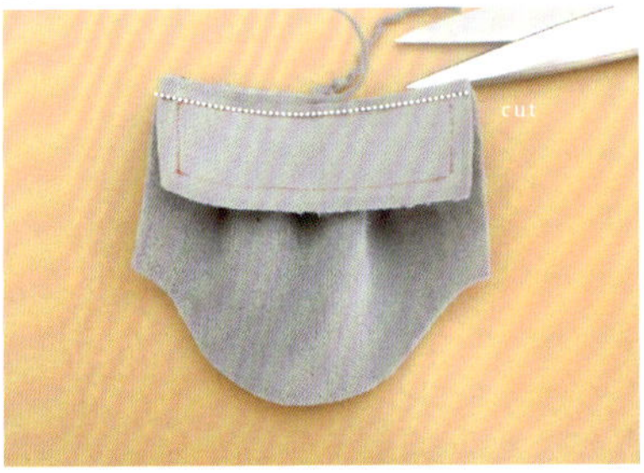
cut

正面向内缝合袖口和袖头，缝份剪成 3mm 宽度。

27

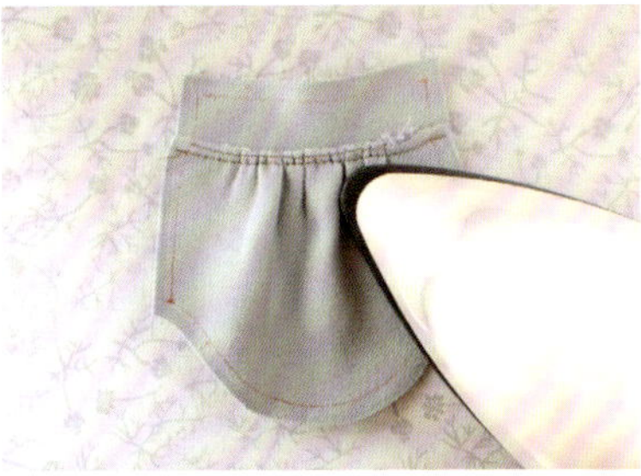

缝份压向袖头侧，翻折袖头。

19. Match the collar to the neck opening and attach with fabric glue. *20.* Make small cuts along the seam of the yoke.
21. Match the yoke and the collar and sew. *22-23.* Fold the armhole of back fabric iron and sew.
24. Gather the sleeve openings until the width fits the cuff. [refer to P.92] [Please jump to P62 image 34-38 in the case of <L> size]
25. Match the sleeve openings to the cuff and sew. *26.* Cut the seam to 3mm. *27.* Iron seam up.

28

熨烫三折边袖头，包住缝份。

29

缭缝固定袖头的底边。

30

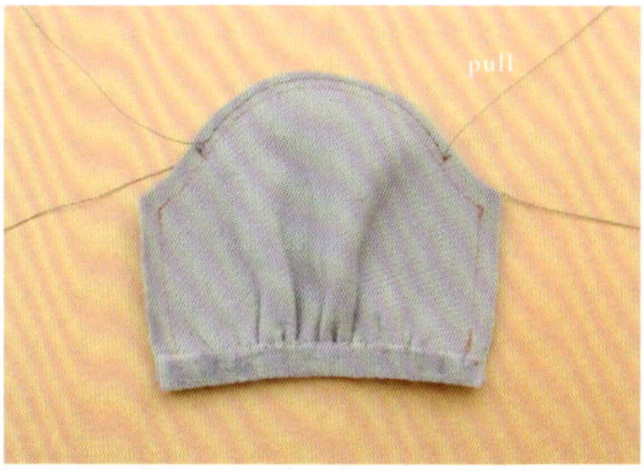
pull

从袖山的缝份的拼合标记至拼合标记，针脚调整为约3mm宽度，单线缩褶车缝。

31

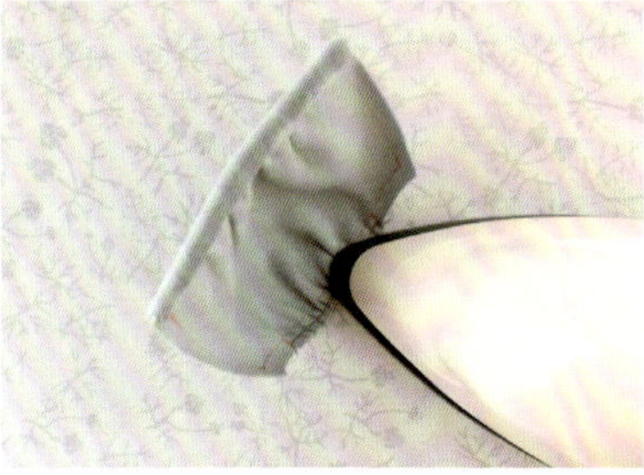

对齐衣片袖窿的宽度进行缩褶，熨烫整齐。

32

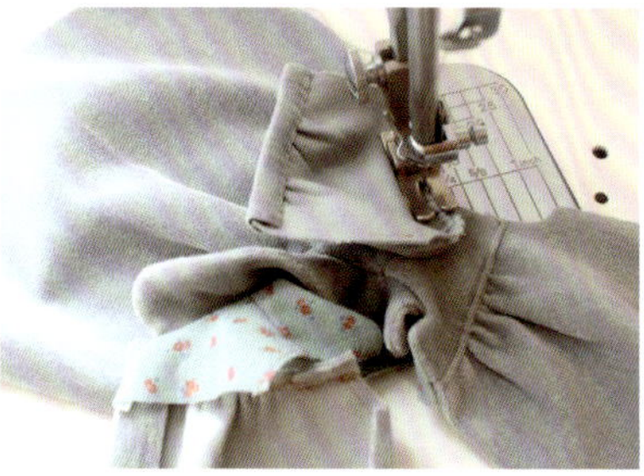

正面对合袖子和衣片。缝合过程中分几次抬升车缝机的压脚，逐渐对齐袖上的缝份和袖窿的缝份。

33

袖子已缝接于衣片。缝份压向袖子侧，熨烫整齐。

34

里衬前衣片的缝份折向内侧，熨烫压实。

35

里衬后衣片的缝份折向内侧，熨烫压实。

36

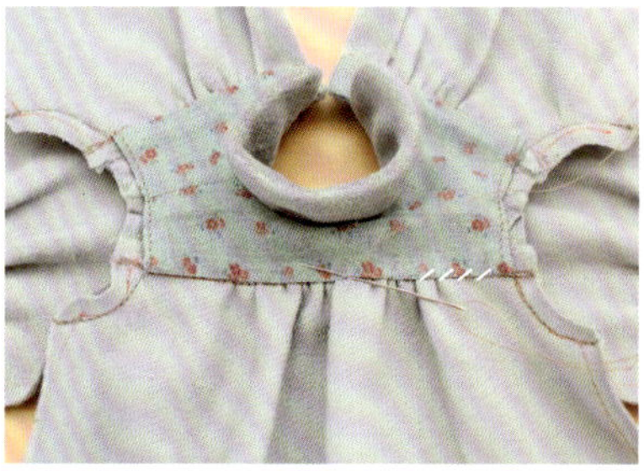

里衬缭缝于后衣片。

28. Fold the cuff around the gathering. *29.* Finish with a blind stitch. *30-31.* Gather the shoulders until the width fits the armhole and iron. *32.* Match the side edge of the sleeve to the bodice and gradually sew the shoulder of the sleeve to the armhole, matching as you go. *33.* Now your sleeves are attached! *34-35.* Fold the edge of the yoke under and iron. *36-37.* Finish with a blind stitch.

37

缭缝里衬和前衣片。

38

正面对合前衣片和后衣片，缝合袖口～侧身～下摆。

39

剪开

侧身的缝份加入剪口。

40

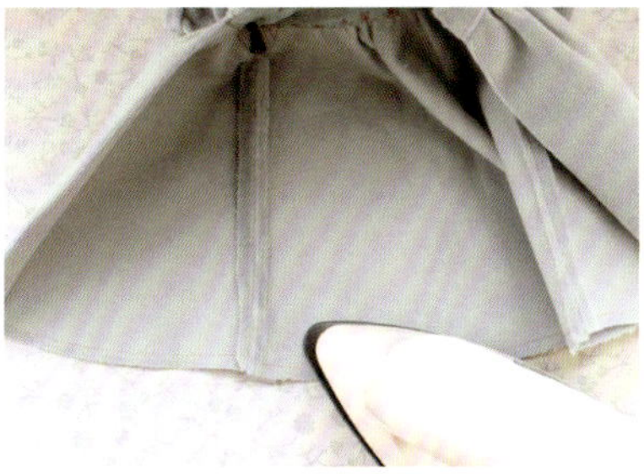

翻到正面，熨烫摊开缝份。

41

剪开

如图所示，裁剪前贴边缝份的下摆。

42

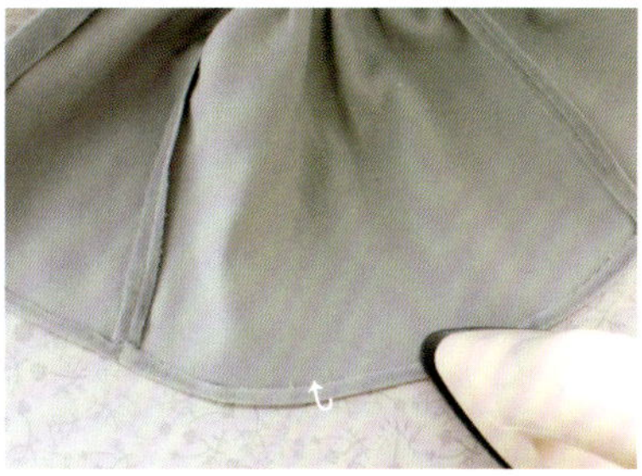

将下摆的缝份熨烫折入内侧。

43

颈部周围、前开衩、下摆、前开襟、颈部周围，整周明线车缝。

44

缝接装饰纽扣。※L 码将按扣和装饰纽扣缝接于袖口。

45

前开衩缝接弹簧扣，另一侧制作线袢（参照第 91 页）。

38. Pin the sides of the bodice and sleeves and sew. *39.* Make cuts in seam allowance of the pits.
40. Turn right side out. Unfold the seams and iron. *41.* Cut the end of the hem as pictured. *42.* Iron the hem and sew.
43. Sew around the edges as pictured. *44.* Attach the button. *45.* Attach the hooks and make thread loops. [refer to P.91]

Shoulder Bag
单肩包

简洁，但能打造外出效果。
厚皮革难以缝制，关键是使用薄且柔软的皮革制作。

本体皮革	S	5cm×4cm	肩带皮革	S	3mm×15.5cm
	M	8cm×6cm		M	3mm×19cm
	L	10cm×7cm		L	3mm×32cm

4mm 圆环　2个

1

对照纸型裁剪。正面相对折叠，缝合两端。车缝缝合时，铺设薄纸一起缝合。

2

缝制始端及末端回针缝，线头打结。

3

翻到正面。

4

胶

包盖的粘贴部分涂抹皮革胶。

5

包盖粘贴于包本体的反面。

6

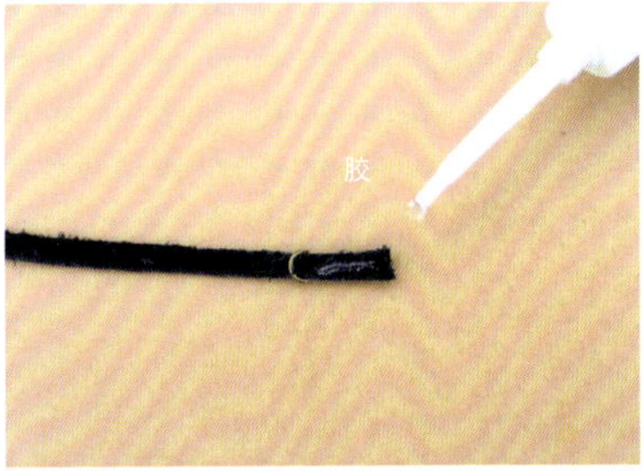
胶

将 4mm 圆环穿入裁剪用于肩带的皮革内，距离端部 1cm 左右涂胶。

7

翻折肩带贴合，待其干燥。

8

圆环缝接于包。

9

单肩包制作完成。

It's so easy to make, but has a lot of impact on any outfit.
A soft, thin leather is recommended for easier sewing.

{ materials : leather, 4mm round jumpring, leather glue }

1-2. Cut out all the parts. Fold the bag in two and sew the sides with paper under the leather.
3. Turn right side out. *4.* Put the leather glue on the top of the flap. *5.* Attach.
6-7. Thread a round jumpring on to the shoulder strap, fold the end over and attach with leather glue. *8.* Sew the shoulder strap on the bag. *9.* Finished!

Boots & Socks

鞋 & 袜子

足底生辉，
还可制作各种颜色搭配。

双色薄皮革	S 各 5cm×8cm M 各 5.5cm×9cm L 各 20cm×13cm	鞋垫薄棉布	S 3cm×2.5cm M 3cm×3cm L 7cm×7.5cm
鞋底厚皮革	S 5cm×3cm M 6cm×4cm L 11cm×7cm	鞋底针织布	S 8cm×8cm M 10cm×10cm L 20cm×20cm
丝带	S、M 60cm L 80cm	厚纸、双面胶、1.5mm 冲头、金属板、金属锤	

1

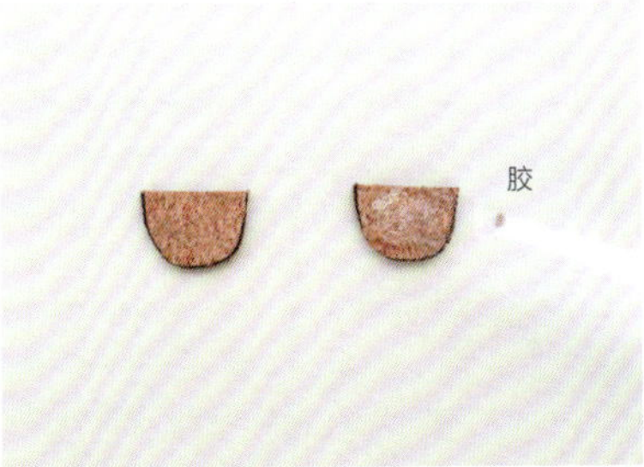
胶

将使用鞋底皮革裁剪的鞋跟部分两片重合，用皮革胶贴合。

2

S 码及 M 码两片一组，L 码四片一组。

3

胶

鞋跟重合于鞋底，用皮革胶贴合。

4

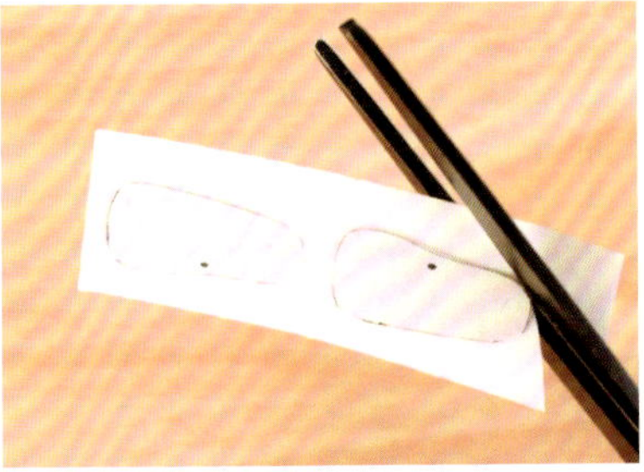

将鞋底的纸型描印于厚纸，制作鞋垫部分，在表面贴双面胶带，并裁剪整齐。

5

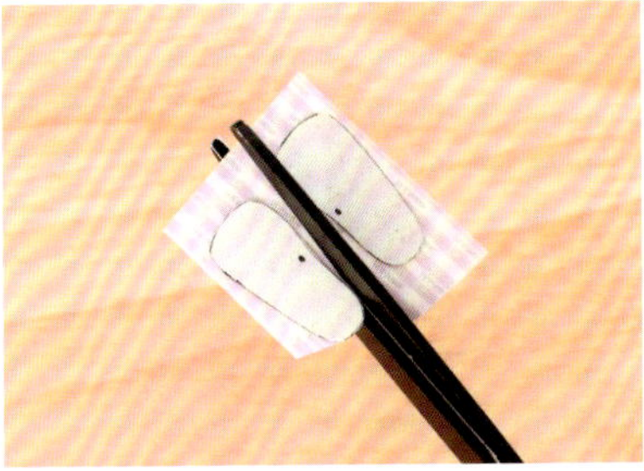

剥开双面胶带的剥离纸，粘贴于鞋垫布料。

6

裁剪鞋垫布料。

7

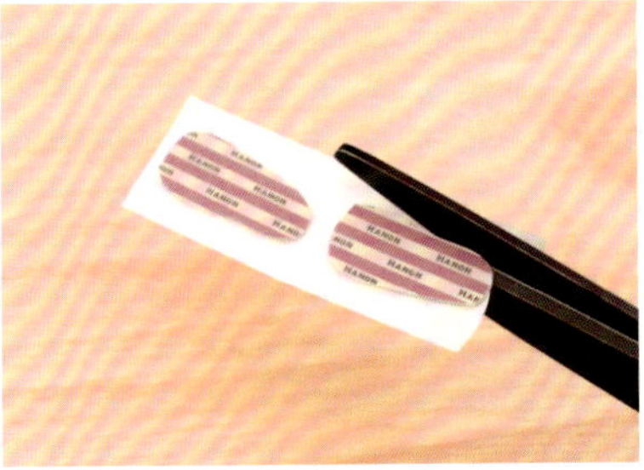

鞋垫反面贴双面胶带，并裁剪。

8

零件 A 的鞋舌周围加入装饰针迹。车缝缝合时，在底部铺设薄纸一起缝合。

9

线头打结处理，剥下纸。

A look starts with a good pair of shoes!!
Try many different colors.

{ materials : leather, cotton, silk ribbon, cardboard, double sided tape, leather glue }

1-2. Glue the heel pieces together, 2pieces for S / M, 4 pieces for L. *3.* Attach the heel to the sole with glue.
4. Trace the insoles on cardboard and cut. Put the cardboard sole on double side tape.
5-6. Attach fabric and cut along the soles. *7.* Attach the other side to double side tape and cut along the soles.
8. Saw along the sides of piece A for decoration, with a piece of paper under A. *9.* Detach the paper slowly so it doesn't tear.

10

零件 B 的穿口周围加入针迹。

11

线头打结处理，剥下纸。

12

对零件 B 开孔，准备 1.5mm 冲头、金属板、金属锤（如果没有冲头，也可使用锥子开孔）。

13

标记使开孔位置左右对齐，冲头垂直于皮革，用锤子敲打冲头开孔。

14

左右边缘分别开孔 4 处。

15

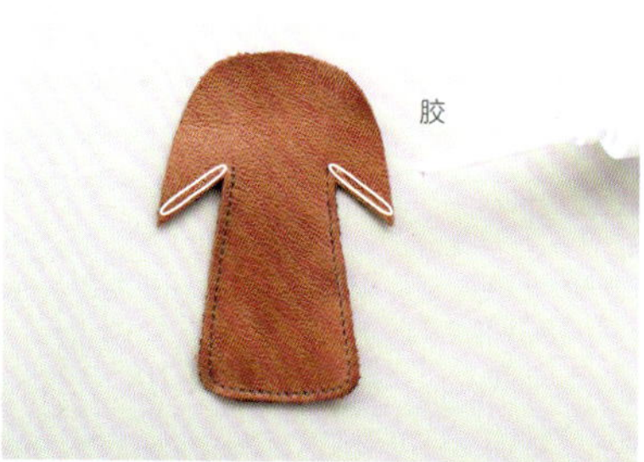
胶

零件 A 表面如图所示位置，涂抹皮革胶。

16

零件 B 的穿口边缘与零件 A 的胶水位置重合 2mm 粘贴。

17

另一侧同样重合粘贴，使胶水完全干燥至粘紧。

18

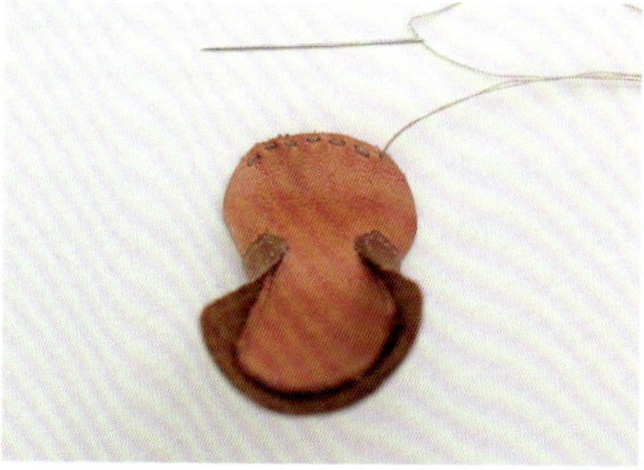

零件 A 的鞋头细密拱缝。

10-11. Repeat with piece B. *12-14.* Use a 1.5mm hole punch and piece part B (for the ribbon).
15. Spread glue along the sides as shown. *16-17.* Attach A and B as shown. *18.* Use gather stitch along the toe.

19

拉

拉线缩褶，制作鞋头的弧线。

20

剥下鞋垫背面的双面胶的剥离纸，嵌入鞋底，包住鞋垫。

21

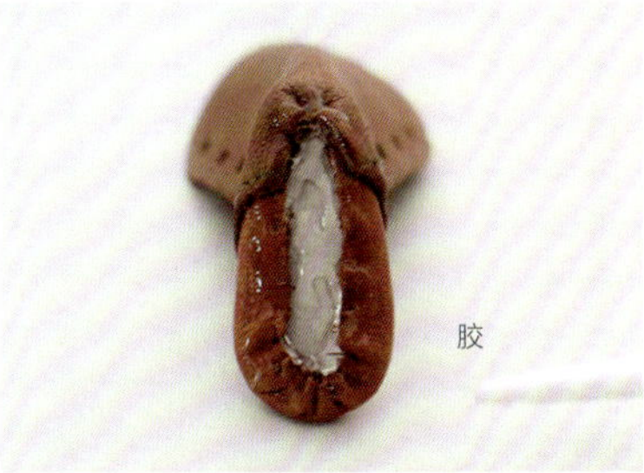
胶

对齐鞋垫将皮革折向内侧，鞋底涂抹皮革胶。

22

粘贴鞋底。

23

胶水充分干燥至粘紧，调整形状。

24

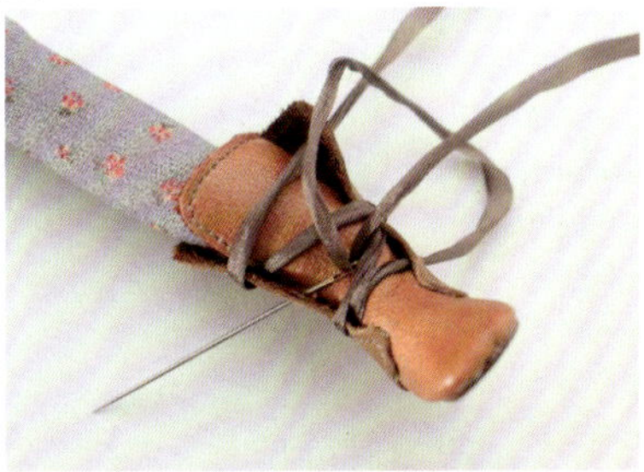

将鞋子穿在娃娃的脚上，穿入丝带。

25

使用针织布制作袜子同样铺上薄纸一起缝合，成型之后更整齐。穿口折向内侧，用针织线缝制。

26

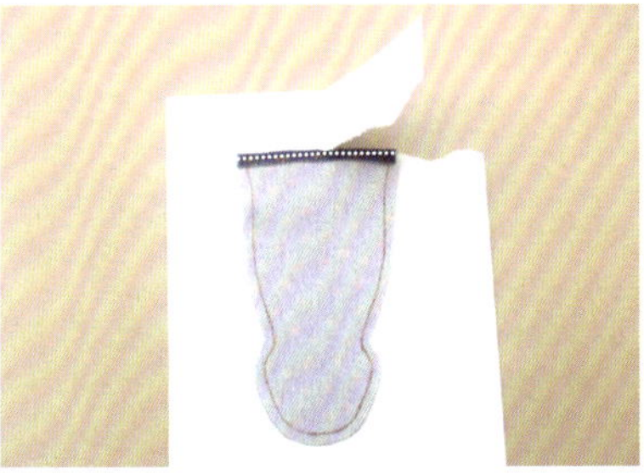

缝制完成后，戳破纸之后拆下布料。

27

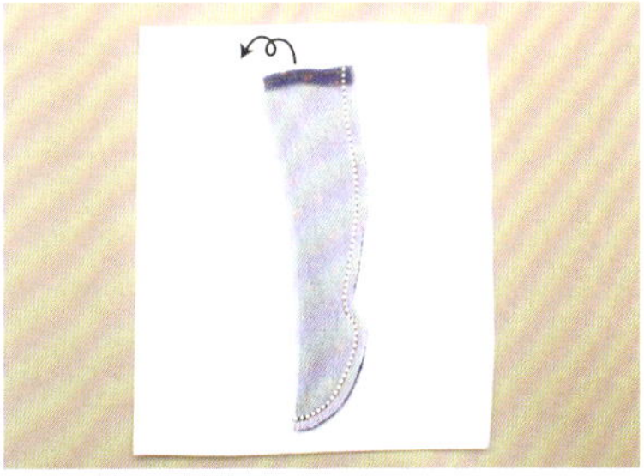

正面向内对折，铺设纸之后缝合。戳破纸，翻到正面即完成。

19. Gather the toe to make it round. *20.* Place the insole inside with the tape side down and stick the leather.
21. Spread leather glue all over the bottom. *22.* Attach the sole. *23.* Once the glue is dry, shape the toe.
24. Thread ribbon through the hole. *25.* Pin and sew the sock opening as shown, keep a piece of paper under the fabric.
26. Tear away the paper. *27.* Fold the sock and sew as shown.Turn right side out.

索罗
雷纳

Sneaky Stuffed Fox

俏皮的狐狸

在外游玩的狐狸。
俏皮的表情让人忍俊不禁。

袖珍皮毛　弟弟（索罗）　15cm×10cm
　　　　　哥哥（雷纳）　15cm×11cm

绣花线　黑色、红色

珠针、丙烯颜料、填充棉

1

ACRYL
GOUACHE
ACRYL
GOUACHE

先制作眼睛。准备黑色及白色的丙烯颜料、画笔。

2

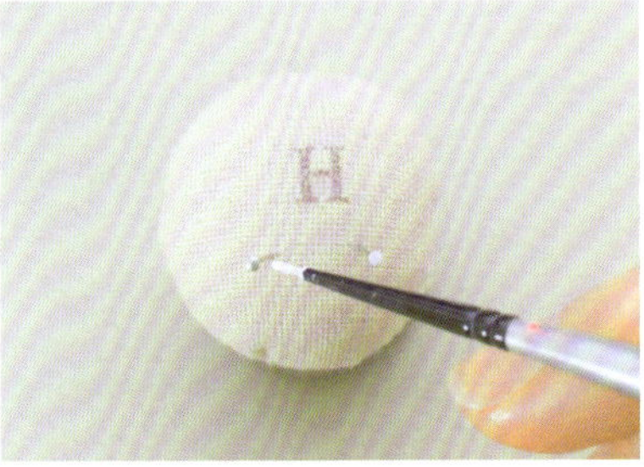

准备 2 个珠针，涂白头部。

3

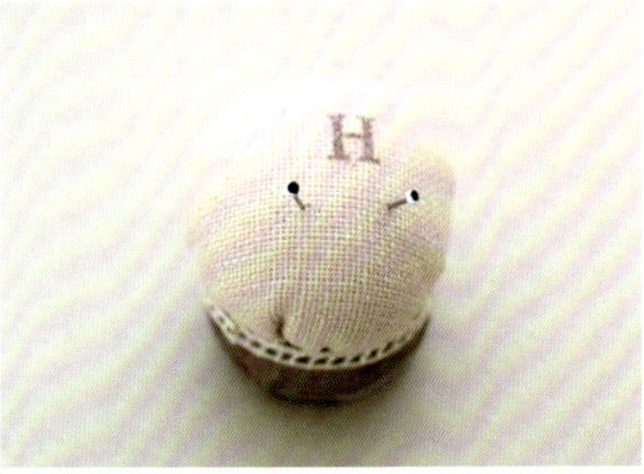

白色干燥之后，描绘黑眼珠。

4

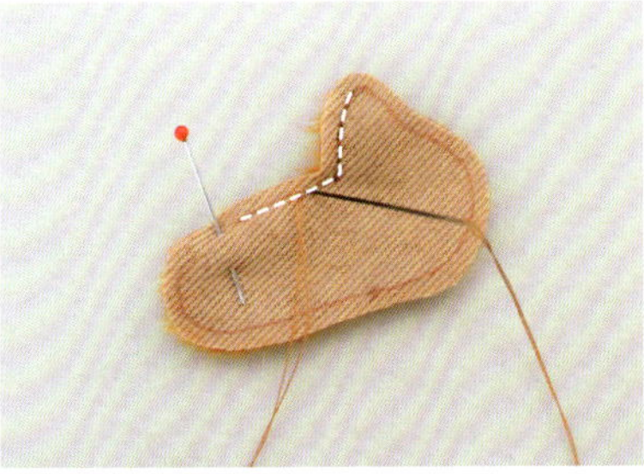

注意布纹的方向，描绘纸型，裁剪各布件。正面对合身体，回针缝鼻下至腹部的开口。

5

对齐拼合标记，将头部的布件送入身体之间，正面对合从鼻尖缝合至后头部。

6

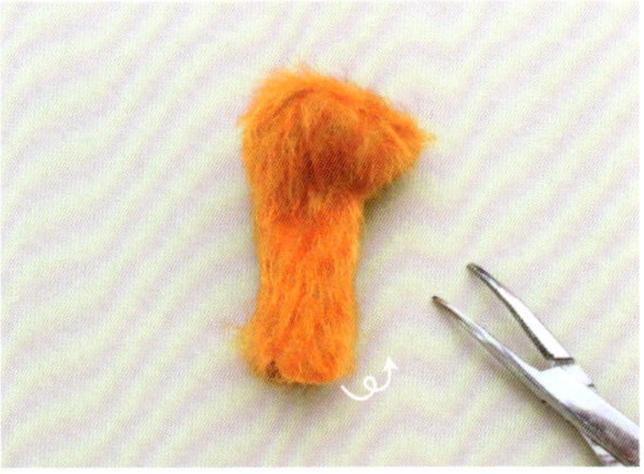

从后头部至翻口正面向内缝合，缝份加入剪口，从下侧的翻口翻到正面。

7

正面对合耳朵、手、脚，留下翻口之后缝合。

8

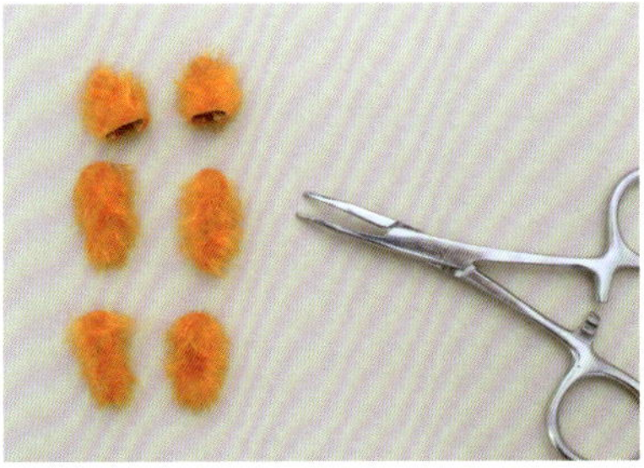

剪掉耳朵的缝份边角，手和脚的缝份加入细密剪口，翻到正面。

9

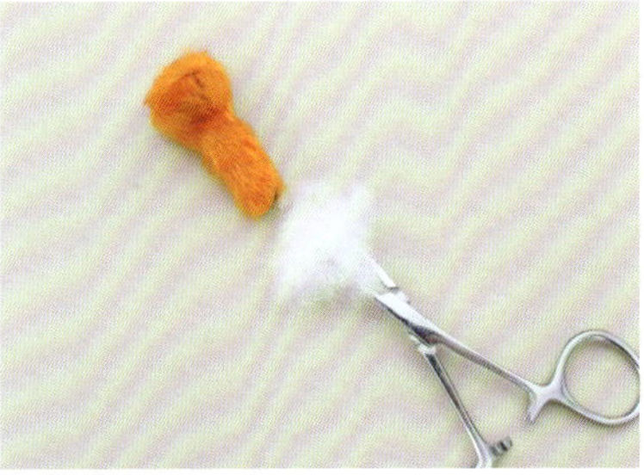

身体塞入填充棉。

Take this sneaky little fox with you out and about.

{ materials: mohair, embroidery thread, pins, acrylic paint, leather glue }

1. For the eyes, use acrylic paint. *2.* Paint the ends of the pins. *3.* When the white is dry, paint black dots. *4.* Match the torso pieces and hand sew starting from the nose as shown. *5.* Sew the top of the head to the torso pieces. *6.* Turn right side out. *7.* Sew the edges of the feet, hands and ears as show. *8.* Turn right side out. *9.* Stuff all the parts.

10

呈 U 字形，缝合返口。手、脚同样塞入填充棉，缝合翻口。

11

手缝接于身体。穿针使线的位置重合于一点，往返多次止缝。

12

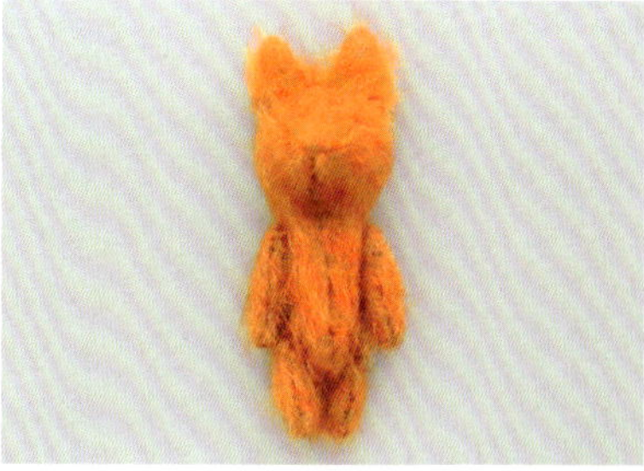

同样方式缝合脚，且确保脚能够运动。

13

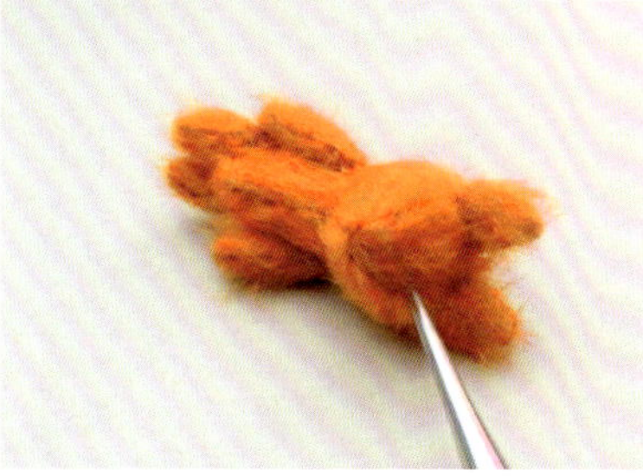

用锥子在眼睛位置开孔，在珠针的眼侧涂胶之后插入。

14

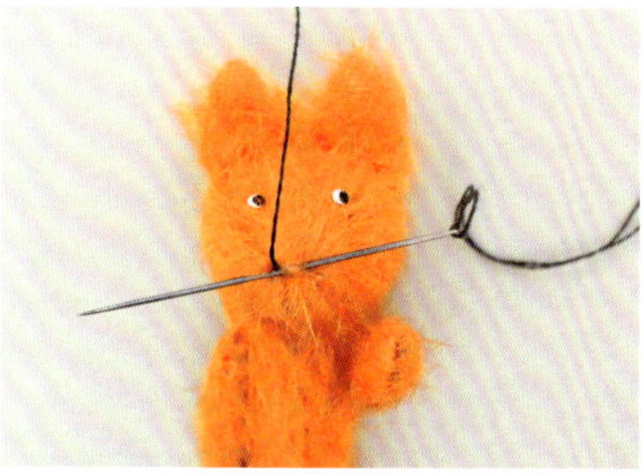

穿 1 根黑色绣花线，鼻子明线车缝。

15

用绣花线制作胡须。距离顶端 2cm 左右位置打结，另一侧打结，线留一点之后剪断。

16

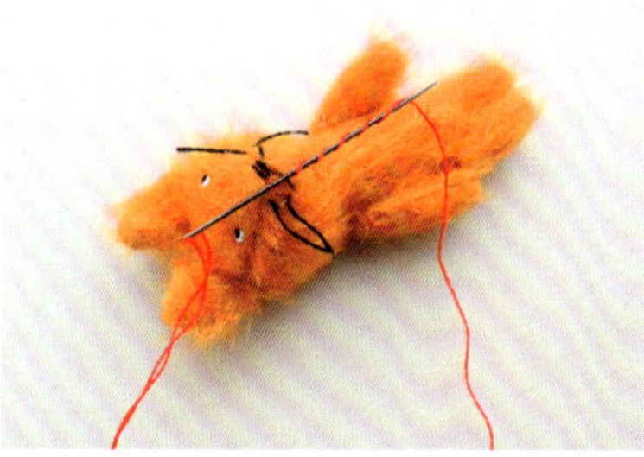

穿 1 根红色绣花线，从需要制作舌头的位置出针，绕绕于针 5 圈。

17

从出针相邻位置入针，拉线完成舌头。

18

丝带固定于颈部，完成。

10. Finish with blind stitch. *11.* Attach the arms by the shoulders going through the torso 3 times. *12.* Attach feet and ears. *13.* Piece the face where you want the eyes, put glue in the holes an insert the pins. *14.* With black embroidery thread, stitch the nose. *15.* Add mouth and whiskers. *16-17.* With red thread, wrap the needle 5 time and stitch to the mouth. *18.* Finish with ribbon!

Gather

缩褶

裙片、袖子、袖头等位置制作缩褶的方法。

1

调节缝纫机的刻度，将针脚的宽度设定为 2.5mm~3mm。

2

缝制始端及末端不回针缝，缝合缝份的正中央位置。

3

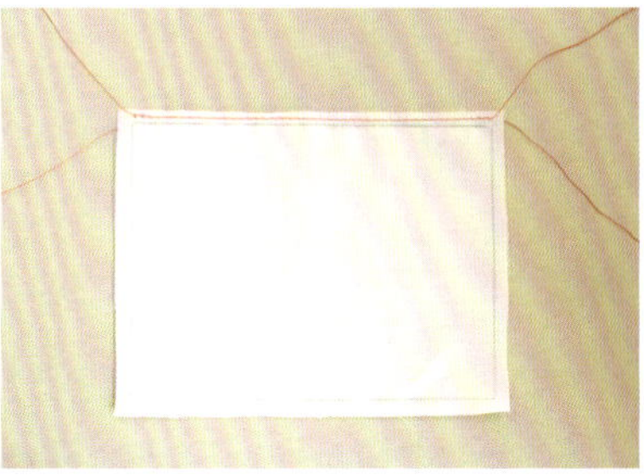

两端的线留下15cm左右，使其容易拉收。

4

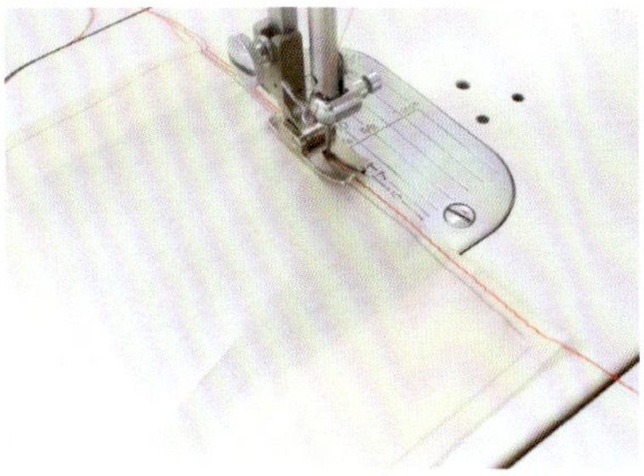

第 1 根线的缝合位置旁边，平行缝合第 2 根线。

5

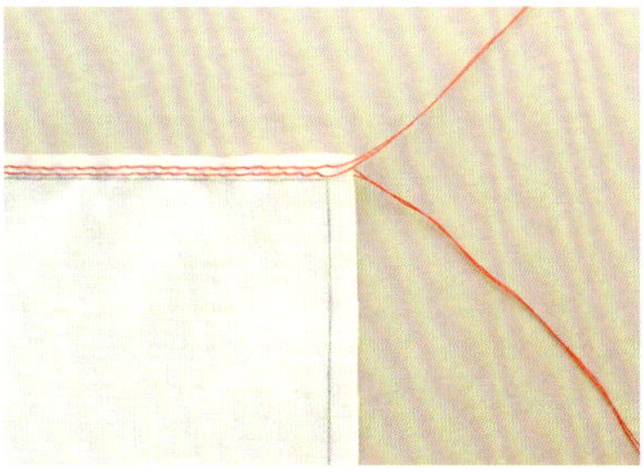

将面线及里线分开。

6

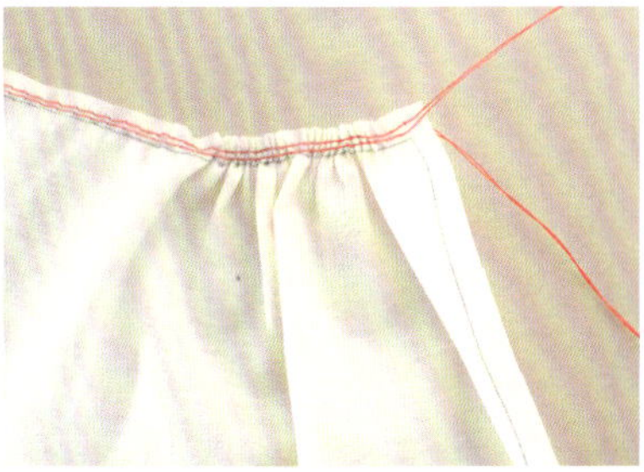

仅拉动 2 根面线，进行缩褶。缩褶距离长则从两侧同时拉动，短则一侧打结，从另一侧拉动。

7

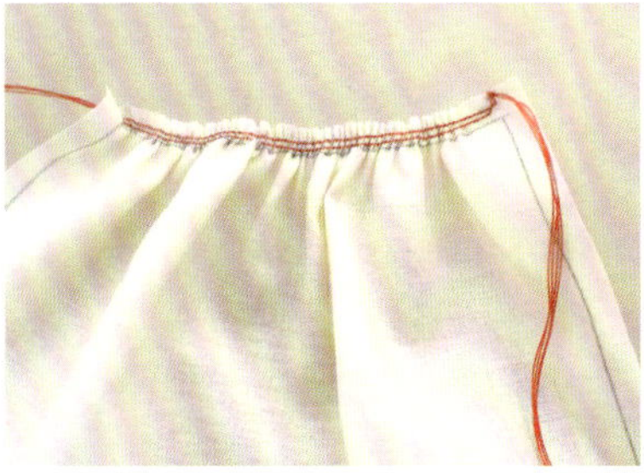

缩褶至所需长度之后，各面线打结，各里线同样打结。另一侧同样打结，固定缩褶宽度。

8

调整缩褶的间隔，熨烫压实。

9

缩褶完成。如果介意缝份的缩褶线，缩褶完成之后也可抽掉。

1. Set the machine to 2.5-3.0mm stitch. *2.* Do not use back stitch as usual the start or end. Sew once along the edge. *3.* Leave about 15cm of thread allowance on each side. *4.* Sew a second line next to the first in the same way. *5.* Separate both upper threads from the lower on each side. *6.* Pull the upper threads while gathering the fabric. *7.* When you have the desired width, knot all threads together on either side. *8.* Iron the gathering to make it neat. *9.* Cut away the thread allowance.

Thread Loop

线袢

背心、外套的前开襟等所需“线袢”的制作方法。

1

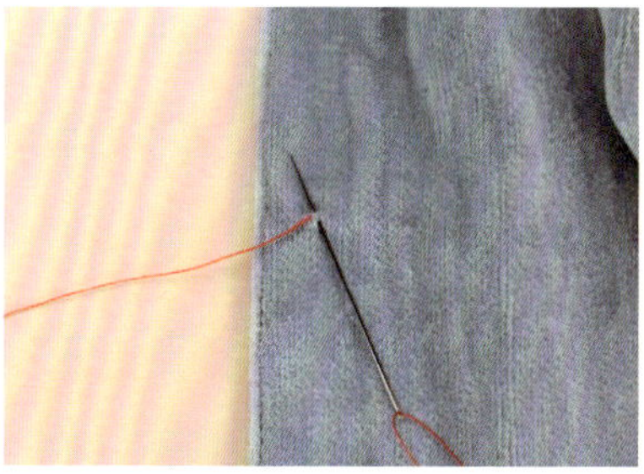

针穿 1 根手缝线。从与弹簧扣重合位置出针，挑起侧面相邻的布料。

2

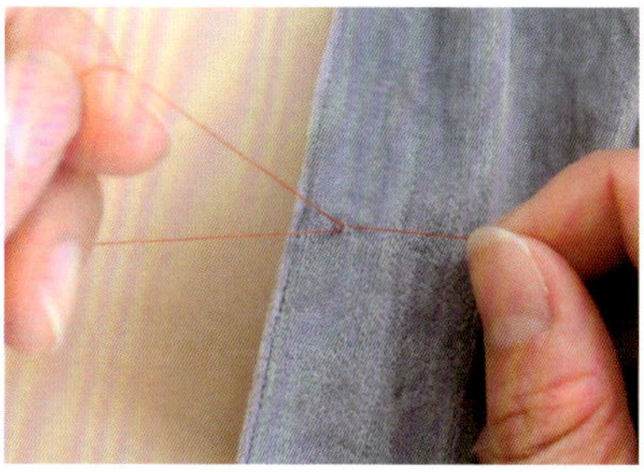

拉线至中间，如图所示制作线环。由此开始用手编入。

3

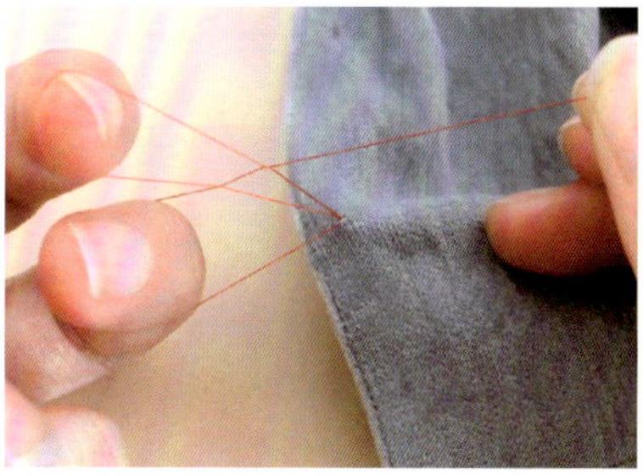

右手的线绕于线环，制作新的线环。

4

拉线，收紧缩小之前的线环。

5

至此编完一针。再将右手的线绕于新的线环，重复增加针圈。

6

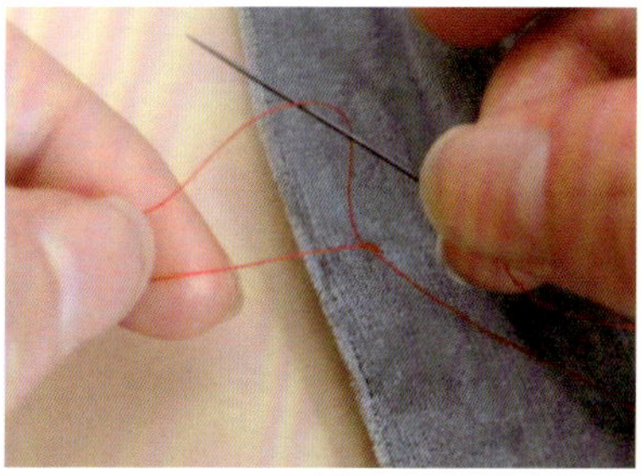

针圈比弹簧扣宽度稍长之后，线穿入线环。

7

拉线，收紧线环。

8

针圈固定。

9

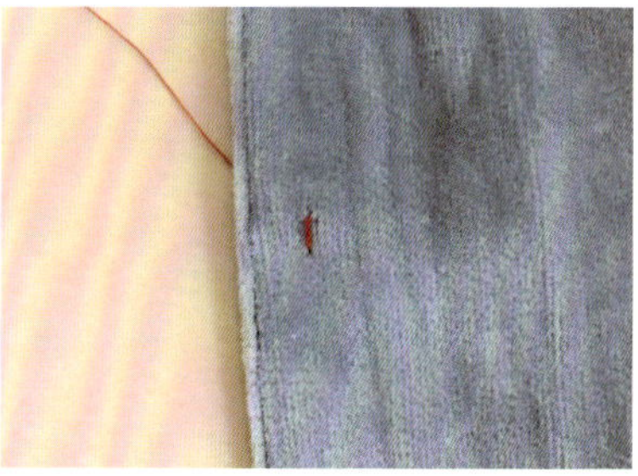

平行于前开襟入针，打结固定，线袢完成。

1. Pull the thread through once. Next to the opening, as if stitching, piece the fabric but don't pull the looped thread all the way.
2. Use your index and ring fingers to grab the loop, and hold the end with your other hand as shown.
3. With your middle finger, grab the end and bring it through the loop.
4-5. Release the thread from your index and ring fingers, and pull the loop on your middle finger.
6. Repeat steps 2-5 until you have the desired length (to match the hook).
7. Thread the needle through the loop and pull. *8.* Fasten the end by pulling the end through the fabric and tie.

Dolls

娃娃

介绍本书中出现的娃娃模特。
除了老式娃娃“Betsy McCall”，许多厂商都有发售各种肤色及发型不同的娃娃。
图片中的模特娃娃由 HANON 实施定制化妆。

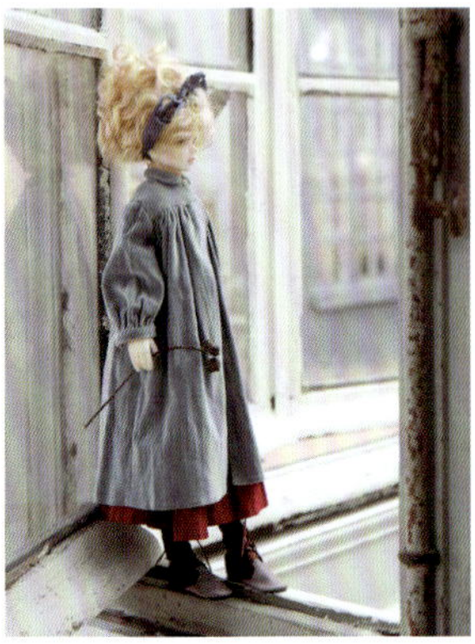

S 尺码

“中布娃娃”

属于小布娃娃和小小布娃娃之间尺寸，由CWC 创作，日本多美玩具公司于 2010 年发售的娃娃。眼睛颜色 1 种，后头部的转盘能够控制眼睛左右转动。全高约 20cm，B8.5×W5.5×H8.0（cm），尺寸相当小。

CWC http://www.blythedoll.com

M 尺码

“小布娃娃”

约三头身，眼睛可变换 4 种颜色的娃娃。1972 年由美国公司最先开始发售，2001 年日本多美玩具公司开始制作。全高约 28.5cm，身体尺寸与“丽佳娃娃”接近，B10.5×W7.5×H10.0（cm）。

CWC http://www.blythedoll.com

L 尺码

“荒木少女”

娃娃造型师荒木元太郎设计的树脂制球体关节娃娃。基本上以未组装、未涂装的套装销售，可自己给娃娃化妆，组装眼睛、头发。由“炼金术工房”定期定制生产，全高约 42cm，B15.5/16.9×W12.5×H18.4（cm）。

炼金术工房 http://www.alchemiclabo.com

小尺寸

“UnoaQuluts Light”

最早由关口玩具发售的 1/6 尺寸的成品娃娃。其特点是可大范围活动关节及自然造型。原型由荒木元太郎制作，全高约 27cm，B10.5×W6.8×H11.0（cm）。

http://www.sekiguchi.co.jp

小尺寸

“Tiny Betsy MaCall”

1957 年由美国公司发售的古董娃娃。2007 年被复制，全高约 20cm，B8.9×W7.2×H10.1（cm）。

※ 目前尚未销售。

DOLL SEWING BOOK

HANON

娃衣缝纫书

纸样

纸样基本为 100% 原尺寸。

S 码的纸样为黑色

M 码的纸样为红色

L 码的纸样为蓝色

按照颜色区分使用，原尺寸描印所需尺码的纸样，整齐裁剪之后使用。

纸样的描印方法

将纸样放在布料反面，使用划分或划粉笔描印粗线“成品线”和外侧的细线“缝份线”。

裁剪时沿着“缝份线”。

对齐粗线“成品线”缝合。

←→	此箭头为布纹的“纵向”朝向（有布边一侧为纵向）。
▶—	此三角标记为“开衩止处”的标记，必须描印。
—	“花边缝接止处”或缩褶位置等制定标记，必须描印。
○～○	表示缩褶的范围。
“制图”	标有“制图”的纸样均在布料上直接制作纸样（裁剪图）。
“左右 × 各 1”	纸样直接放在布料上制作 1 片，纸样翻面之后颠倒左右制作 1 片，合计制作 2 片。
“表里 × 各 1”	纸样直接放在表布上制作 1 片，纸样直接放在里布上制作 1 片，合计制作 2 片。
“×2”	直接将纸样放在布料上制作 2 片。

Lace Strap Dress

蕾丝吊带连衣裙

制作方法：p. 34

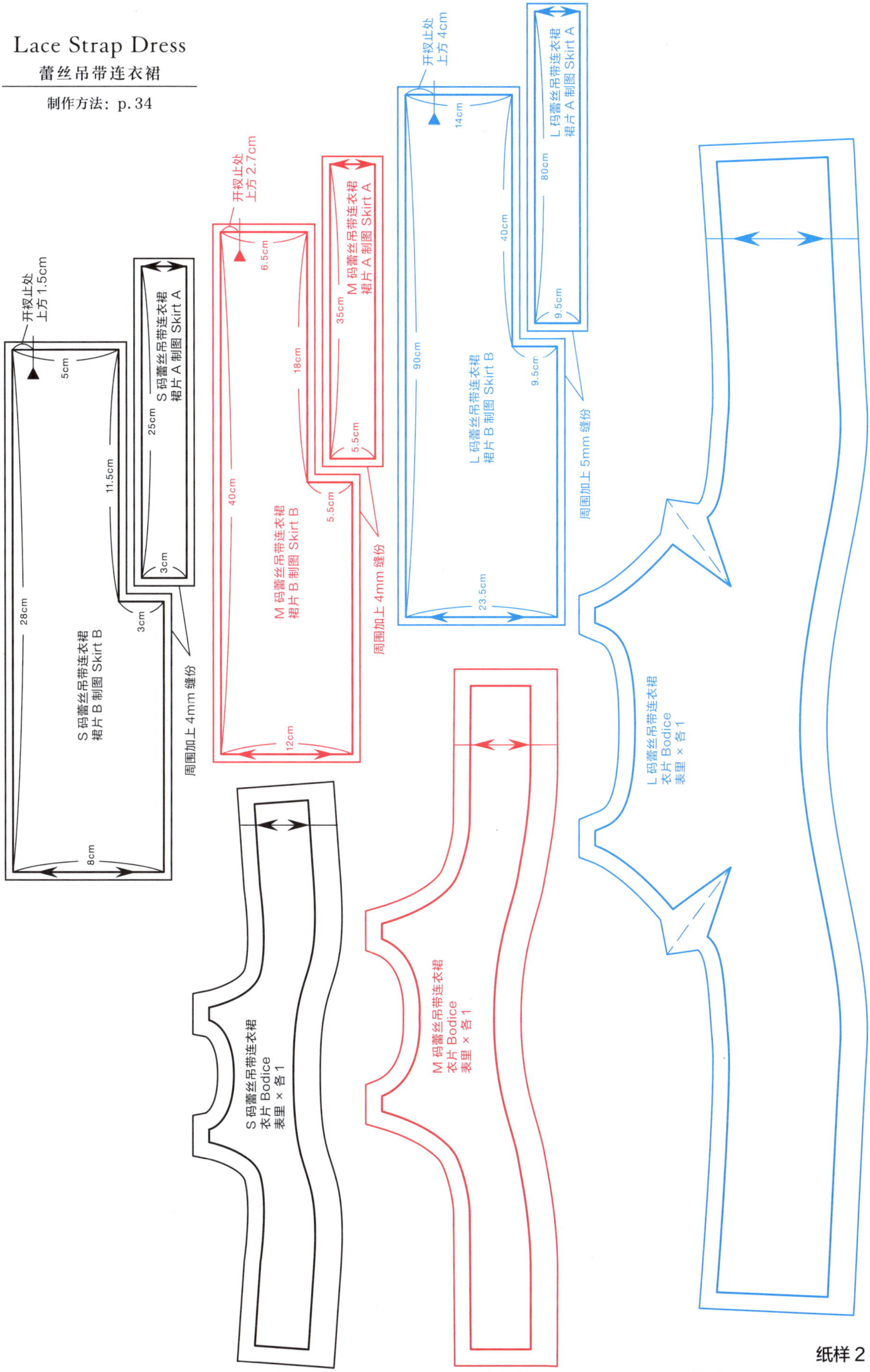

开衩止处
上方 1.5cm
5cm
11.5cm
28cm
3cm
8cm
S 码蕾丝吊带连衣裙
裙片 B 制图 Skirt B
S 码蕾丝吊带连衣裙
裙片 A 制图 Skirt A
25cm
3cm
周围加上 4mm 缝份
开衩止处
上方 2.7cm
6.5cm
18cm
40cm
5.5cm
12cm
M 码蕾丝吊带连衣裙
裙片 B 制图 Skirt B
M 码蕾丝吊带连衣裙
裙片 A 制图 Skirt A
35cm
5.5cm
周围加上 4mm 缝份
开衩止处
上方 4cm
14cm
40cm
90cm
9.5cm
23.5cm
L 码蕾丝吊带连衣裙
裙片 B 制图 Skirt B
L 码蕾丝吊带连衣裙
裙片 A 制图 Skirt A
80cm
9.5cm
周围加上 5mm 缝份
S 码蕾丝吊带连衣裙
衣片 Bodice
表里 × 各 1
M 码蕾丝吊带连衣裙
衣片 Bodice
表里 × 各 1
L 码蕾丝吊带连衣裙
衣片 Bodice
表里 × 各 1

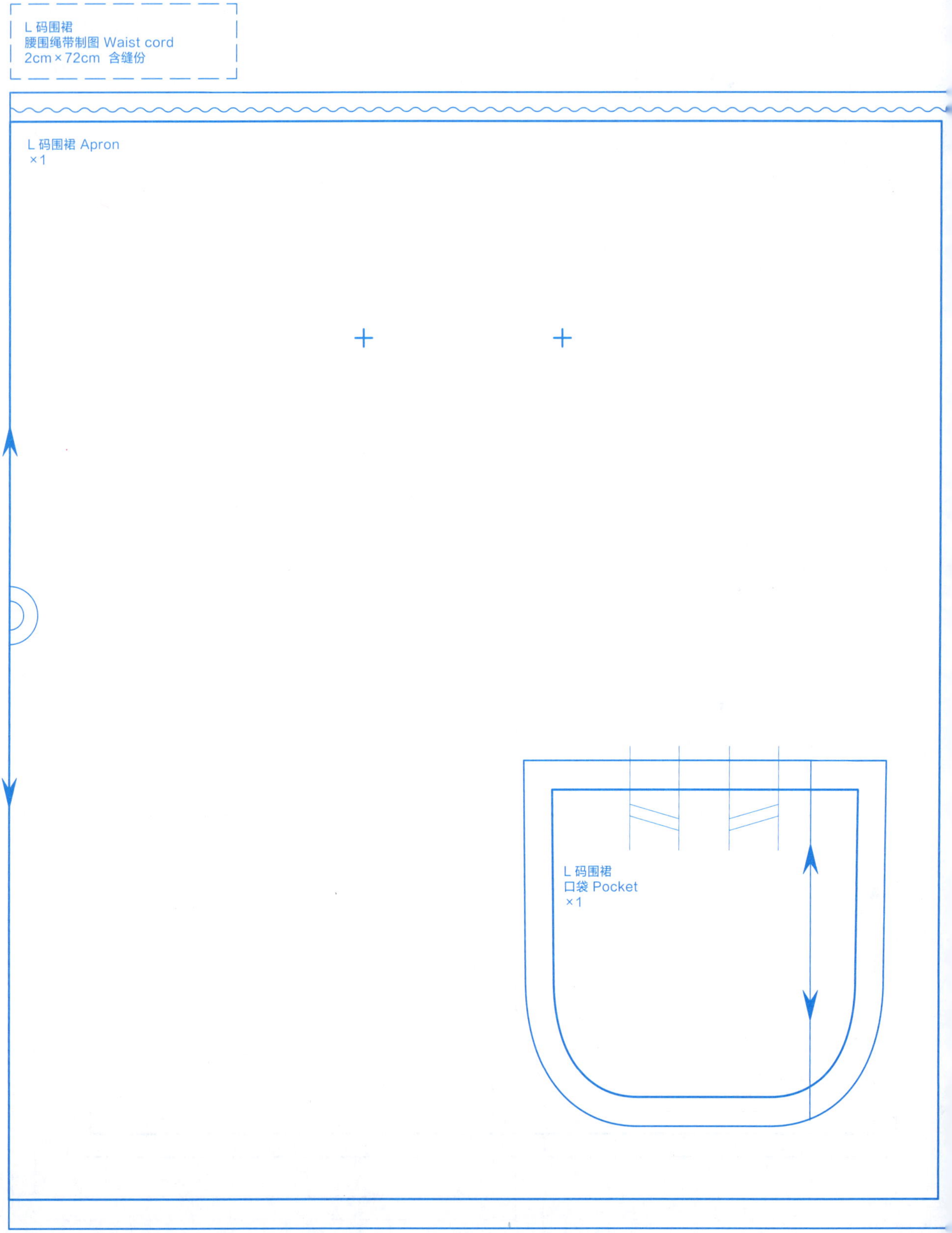
L 码围裙
腰围绳带制图 Waist cord
2cm×72cm 含缝份
L 码围裙 Apron
×1
L 码围裙
口袋 Pocket
×1

Apron
围裙

制作方法：p. 46

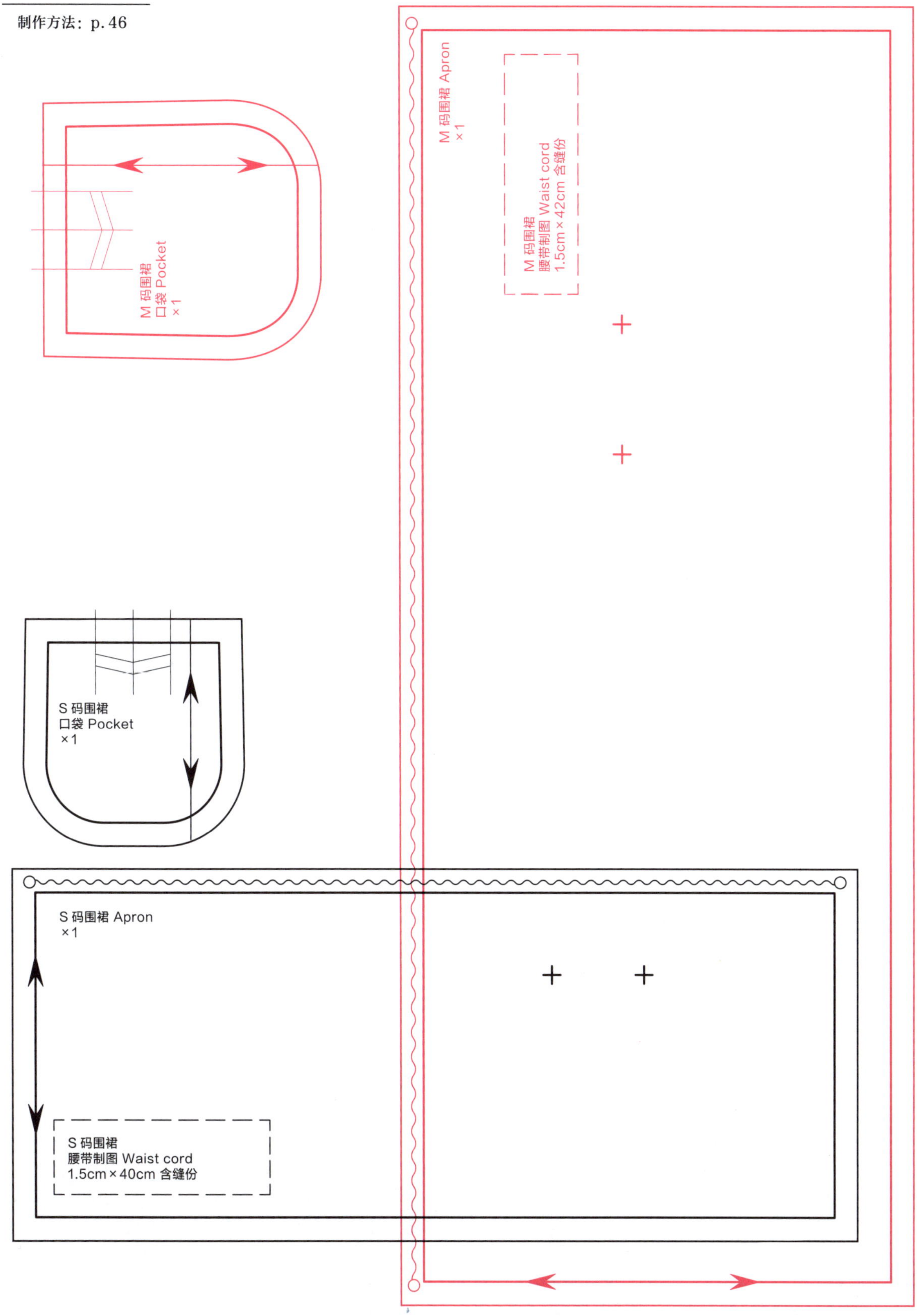
M 码围裙
口袋 Pocket
×1
M 码围裙 Apron
×1
M 码围裙
腰带制图 Waist cord
1.5cm × 42cm 含缝份
S 码围裙
口袋 Pocket
×1
S 码围裙 Apron
×1
S 码围裙
腰带制图 Waist cord
1.5cm × 40cm 含缝份

Vest
背心

制作方法：p.66

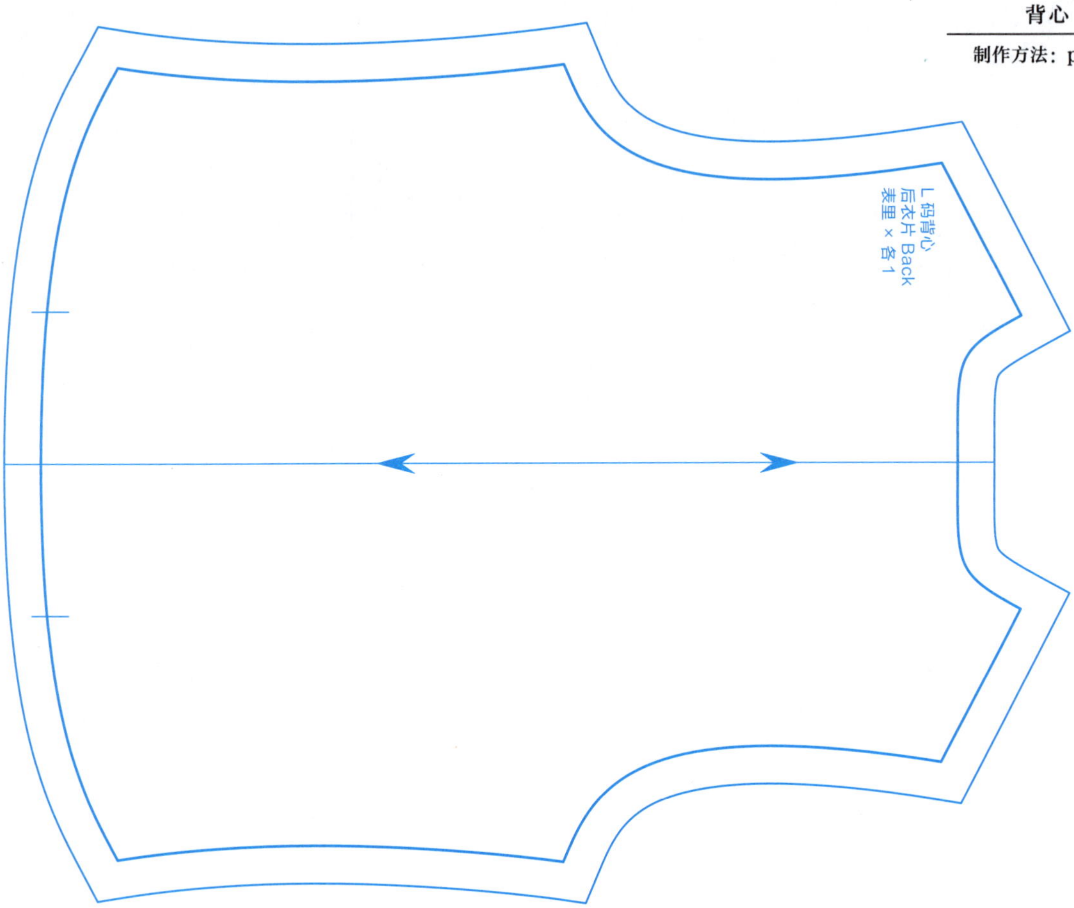
L码背心
后衣片 Back
表里 × 各1

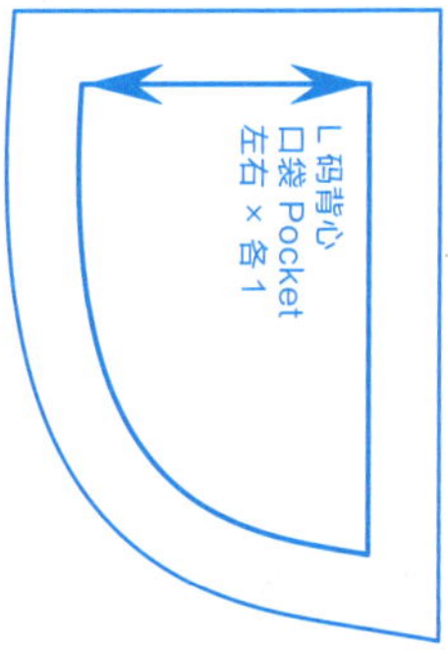
L码背心
口袋 Pocket
左右 × 各1

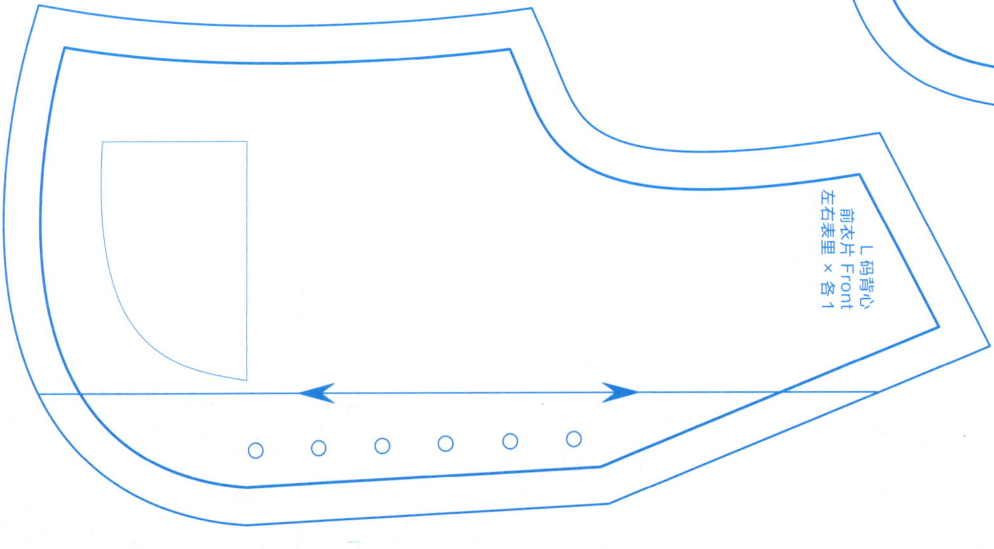
L码背心
前衣片 Front
左右表里 × 各1

Vest

背心

制作方法：p. 66

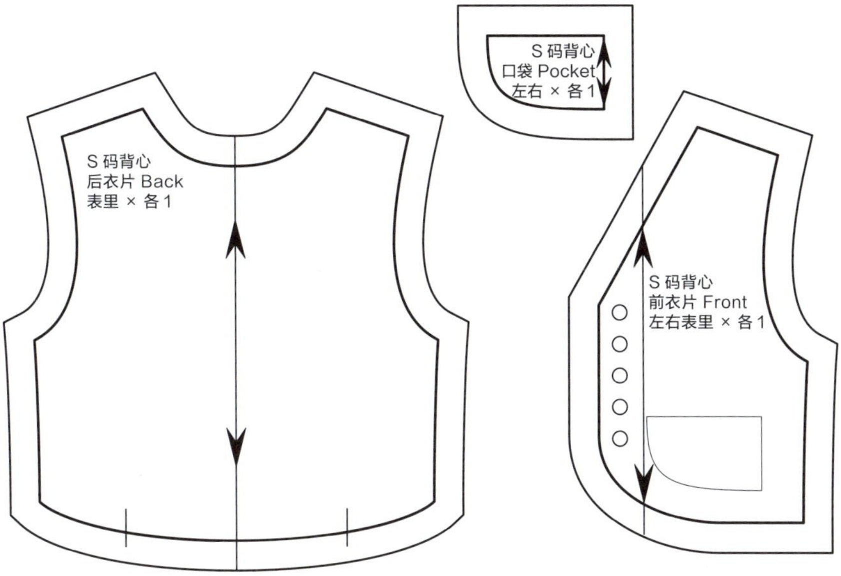
S 码背心
后衣片 Back
表里 × 各 1
S 码背心
口袋 Pocket
左右 × 各 1
S 码背心
前衣片 Front
左右表里 × 各 1

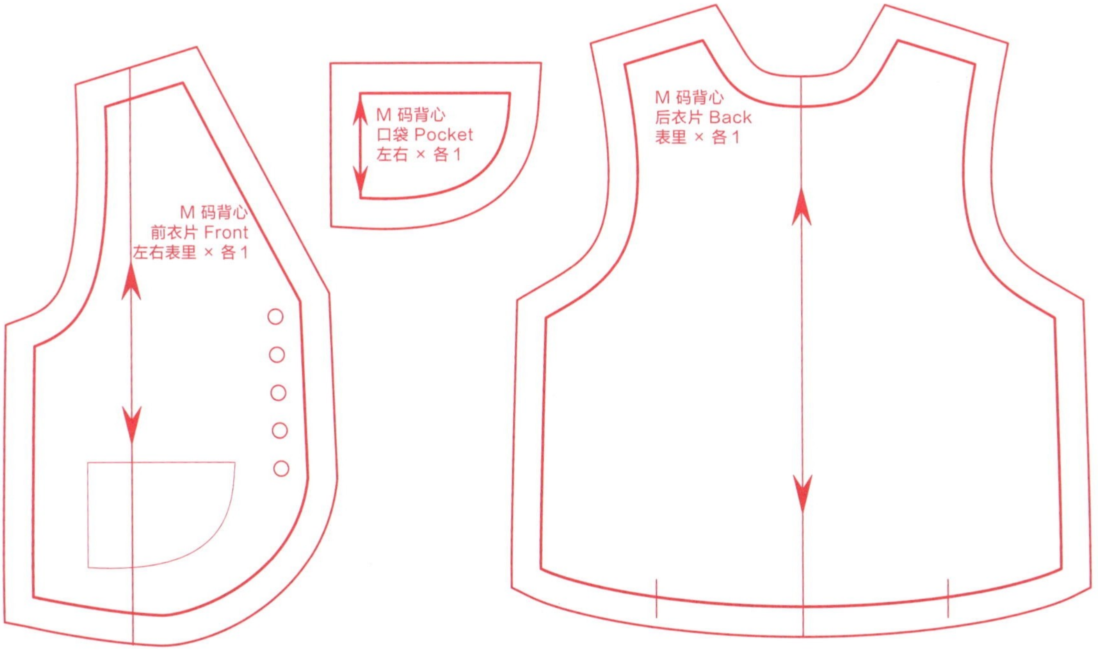
M 码背心
前衣片 Front
左右表里 × 各 1
M 码背心
口袋 Pocket
左右 × 各 1
M 码背心
后衣片 Back
表里 × 各 1

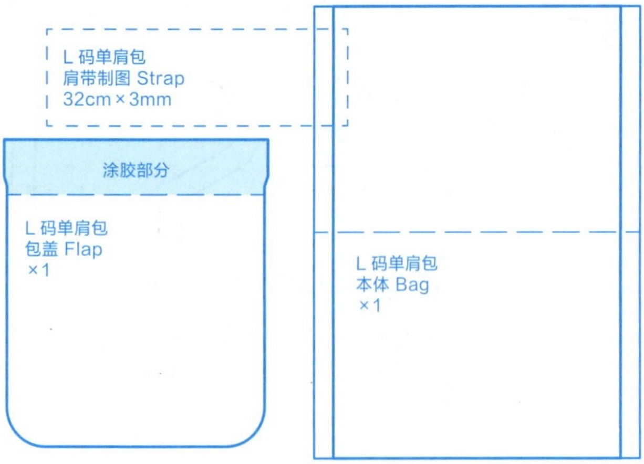
L 码单肩包
肩带制图 Strap
32cm × 3mm
涂胶部分
L 码单肩包
包盖 Flap
×1
L 码单肩包
本体 Bag
×1

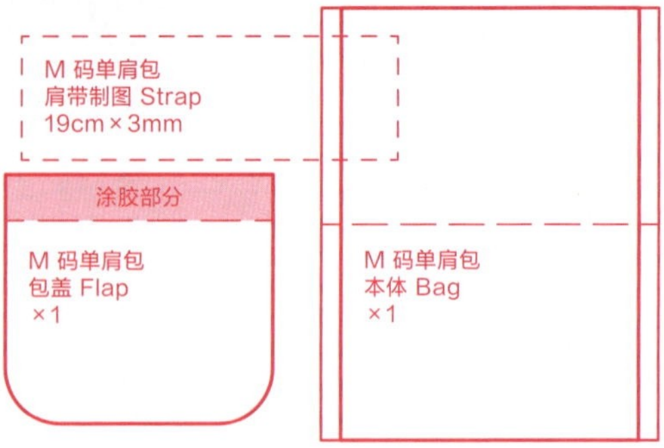
M 码单肩包
肩带制图 Strap
19cm × 3mm
涂胶部分
M 码单肩包
包盖 Flap
×1
M 码单肩包
本体 Bag
×1

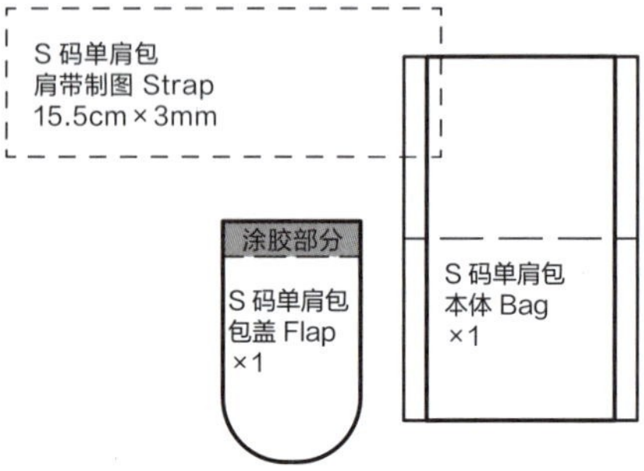
S 码单肩包
肩带制图 Strap
15.5cm × 3mm
涂胶部分
S 码单肩包
包盖 Flap
×1
S 码单肩包
本体 Bag
×1

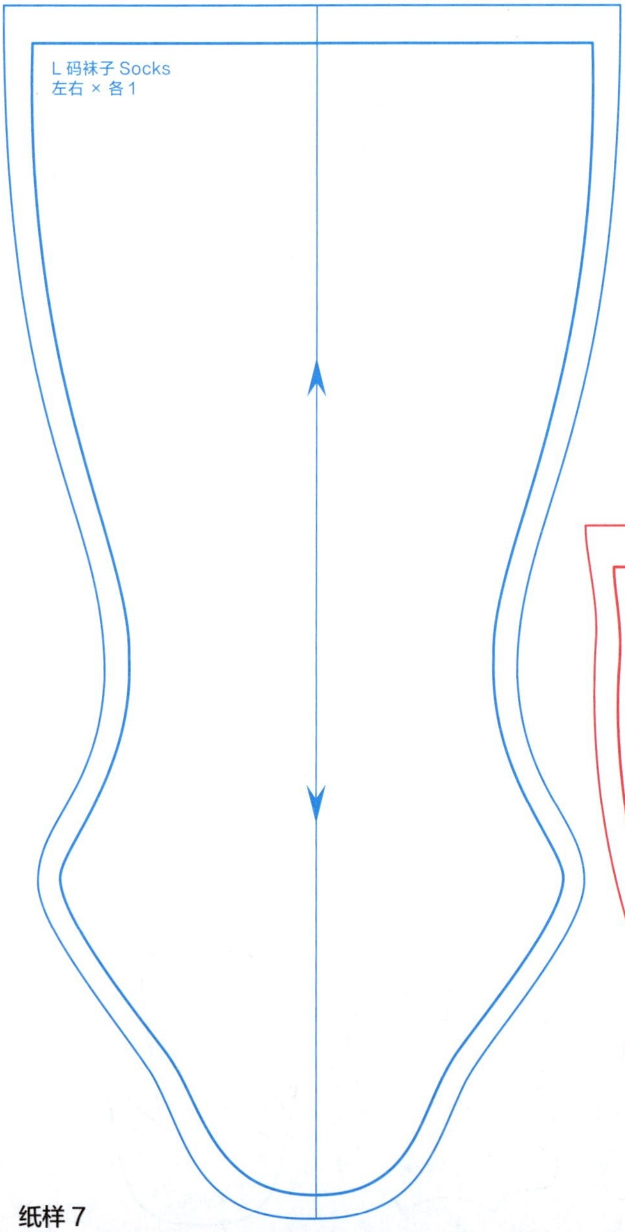
L 码袜子 Socks
左右 × 各 1

Shoulder Bag
单肩包

制作方法：p. 80

Socks
袜子

制作方法：p. 82

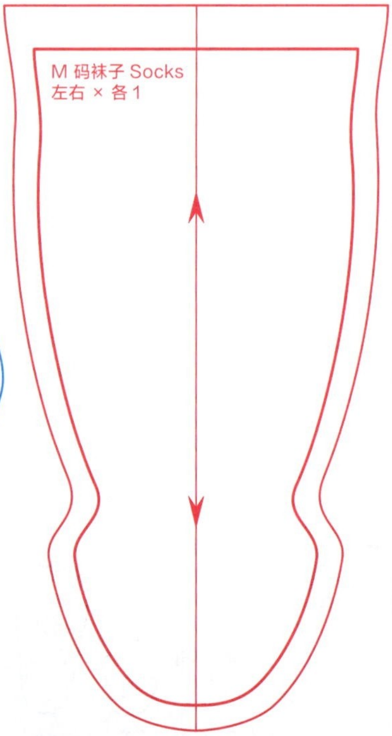
M 码袜子 Socks
左右 × 各 1

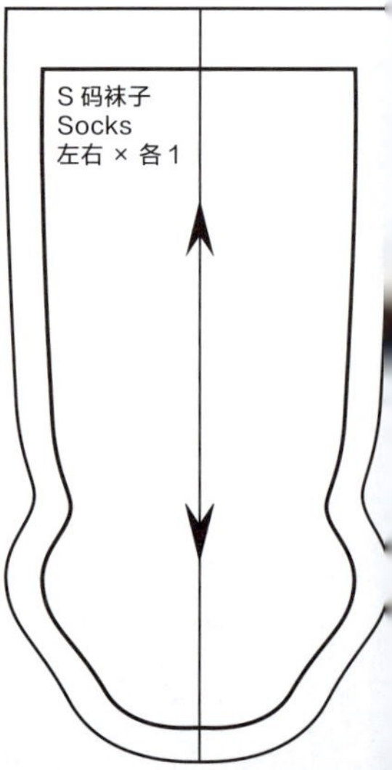
S 码袜子
Socks
左右 × 各 1

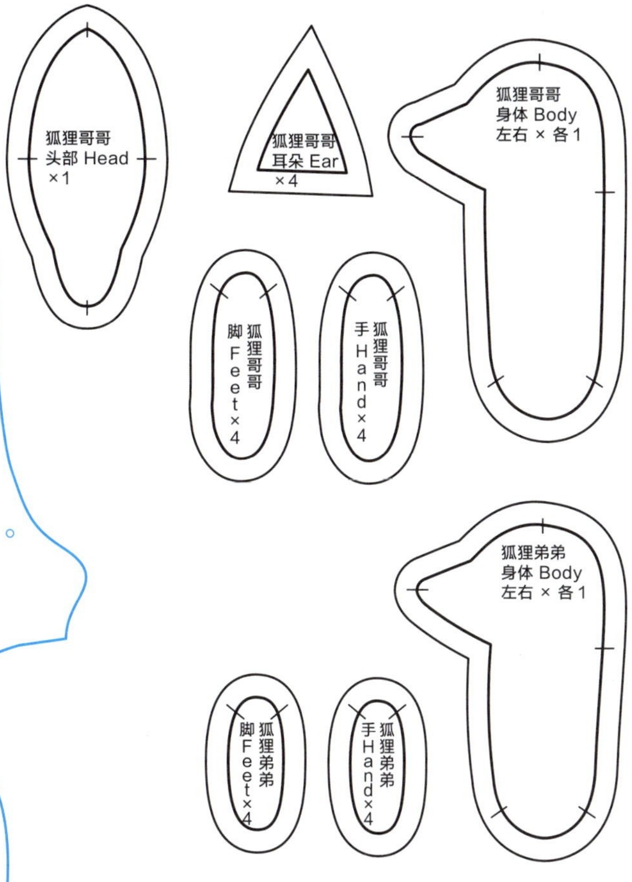
狐狸哥哥
头部 Head
×1
狐狸哥哥
耳朵 Ear
×4
狐狸哥哥
身体 Body
左右 × 各1
狐狸哥哥
脚 Feet×4
狐狸哥哥
手 Hand×4
狐狸弟弟
身体 Body
左右 × 各1
狐狸弟弟
脚 Feet×4
狐狸弟弟
手 Hand×4

Sneaky Stuffed Fox
俏皮的狐狸

制作方法：p. 86

Boots
鞋

制作方法：p. 82

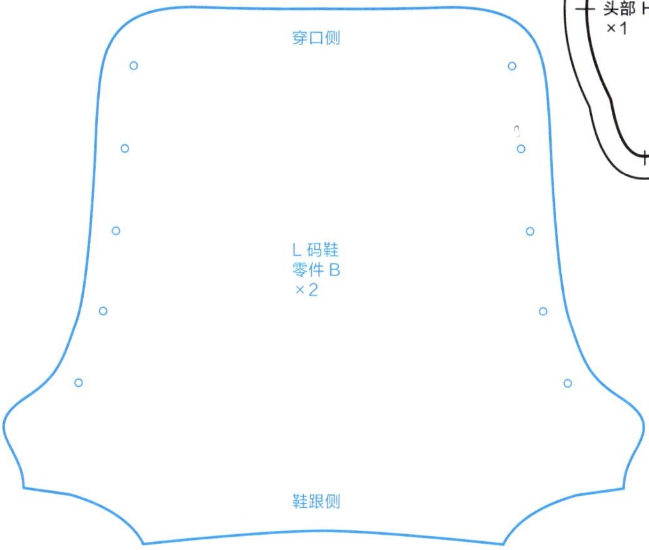
穿口侧
L 码鞋
零件 B
×2
鞋跟侧

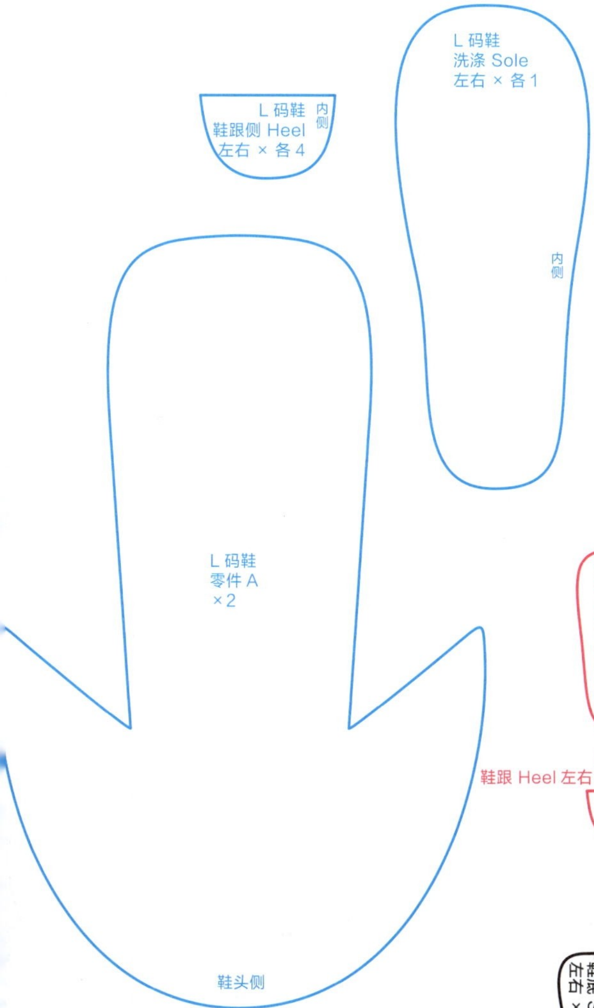
L 码鞋
鞋跟侧 Heel
左右 × 各 4
内侧
L 码鞋
洗涤 Sole
左右 × 各 1
内侧
L 码鞋
零件 A
×2
鞋头侧

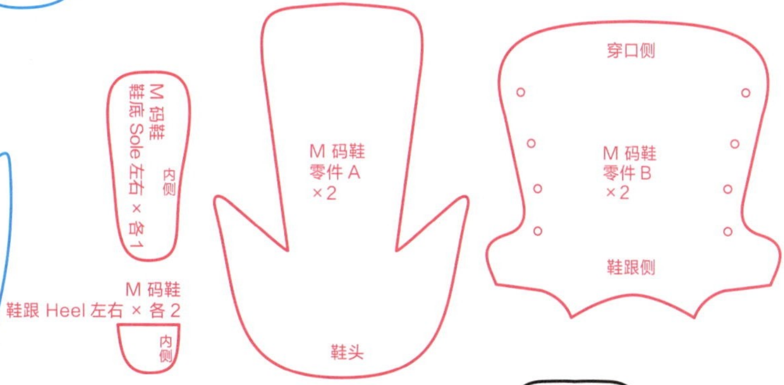
M 码鞋
鞋底 Sole 左右 × 各1
内侧
M 码鞋
鞋跟 Heel 左右 × 各 2
内侧
M 码鞋
零件 A
×2
鞋头
穿口侧
M 码鞋
零件 B
×2
鞋跟侧

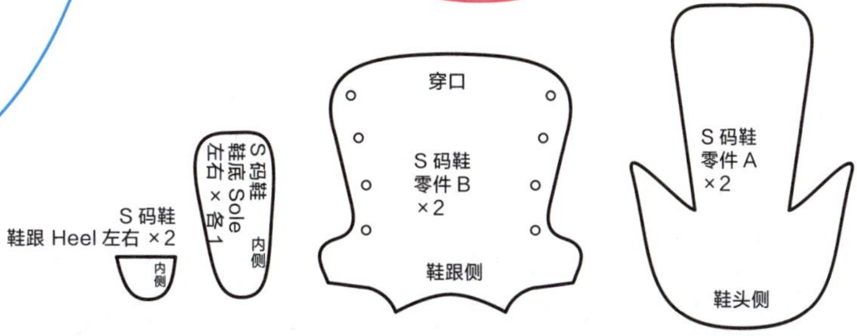
S 码鞋
鞋跟 Heel 左右 ×2
内侧
S 码鞋
鞋底 Sole
左右 × 各1
内侧
穿口
S 码鞋
零件 B
×2
鞋跟侧
S 码鞋
零件 A
×2
鞋头侧

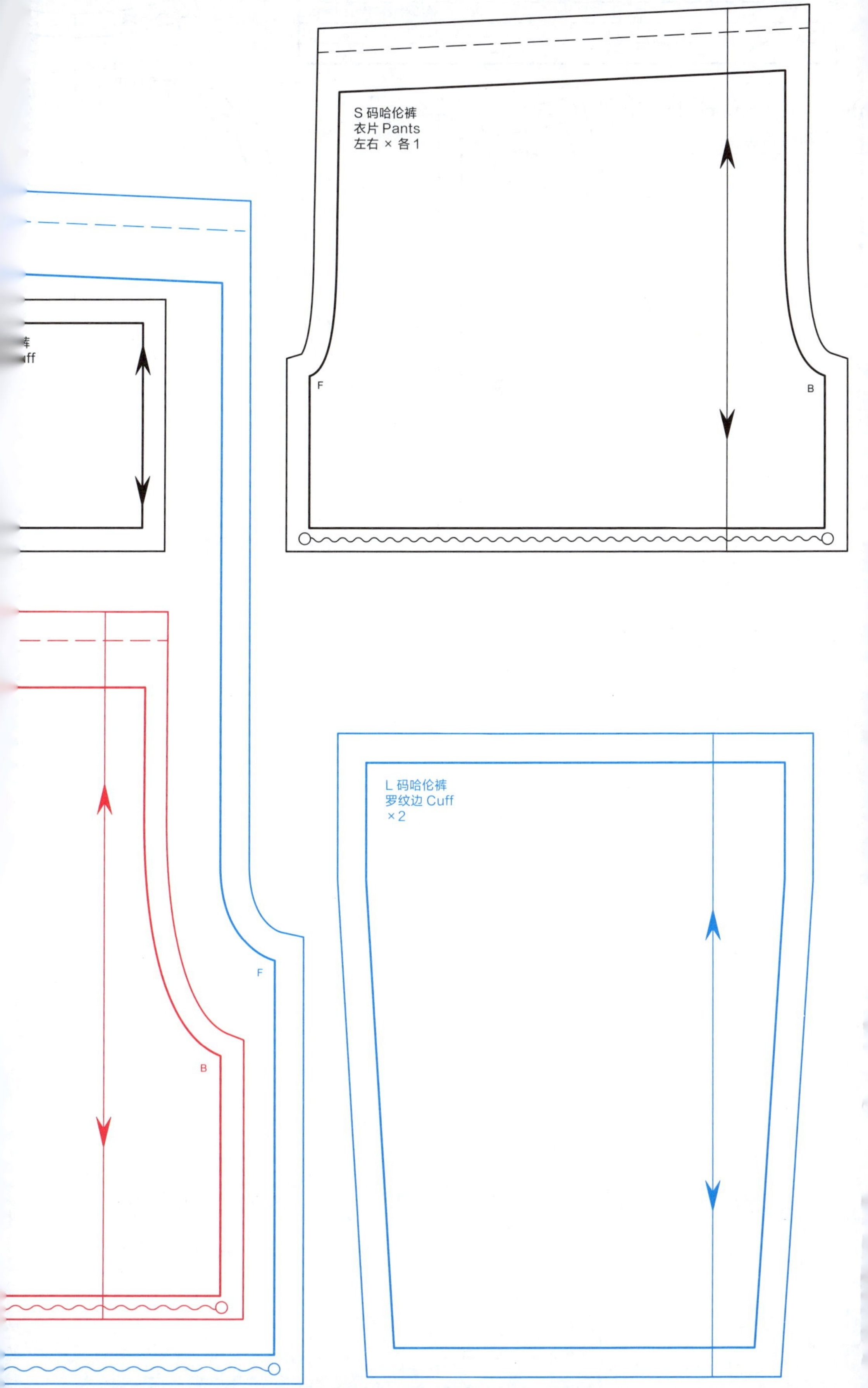
S 码哈伦裤
衣片 Pants
左右 × 各 1
F
B
L 码哈伦裤
罗纹边 Cuff
× 2
F
B

bbroidered Smock Dress

缩褶绣连衣裙

制作方法：p. 42

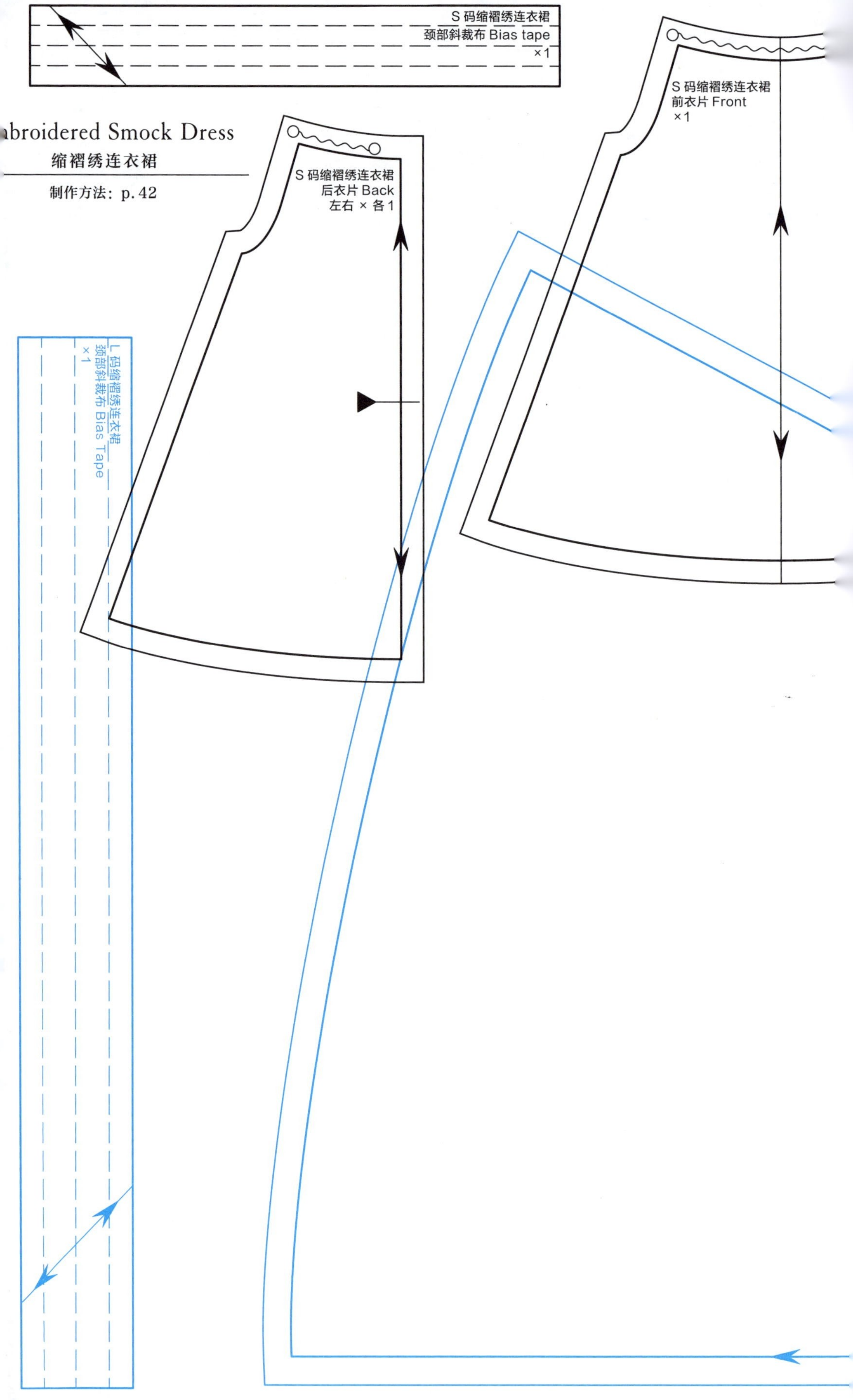
S 码缩褶绣连衣裙
颈部斜裁布 Bias tape
×1
S 码缩褶绣连衣裙
后衣片 Back
左右 × 各 1
S 码缩褶绣连衣裙
前衣片 Front
×1
L 码缩褶绣连衣裙
颈部斜裁布 Bias Tape
×1

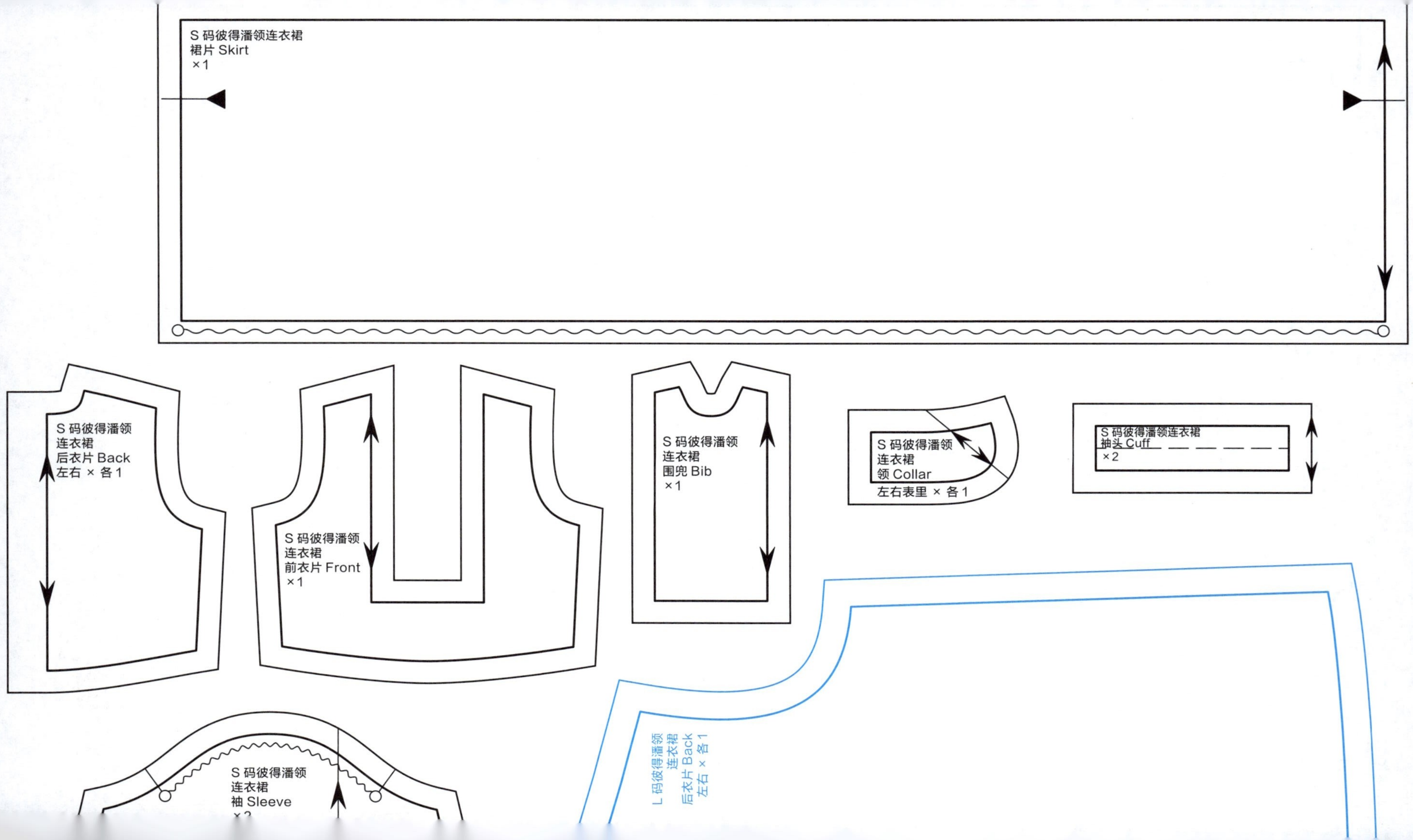
S 码彼得潘领连衣裙
裙片 Skirt
×1
S 码彼得潘领
连衣裙
后衣片 Back
左右 × 各 1
S 码彼得潘领
连衣裙
前衣片 Front
×1
S 码彼得潘领
连衣裙
围兜 Bib
×1
S 码彼得潘领
连衣裙
领 Collar
左右表里 × 各 1
S 码彼得潘领连衣裙
袖头 Cuff
×2
S 码彼得潘领
连衣裙
袖 Sleeve
L 码彼得潘领
连衣裙
后衣片 Back
左右 × 各 1

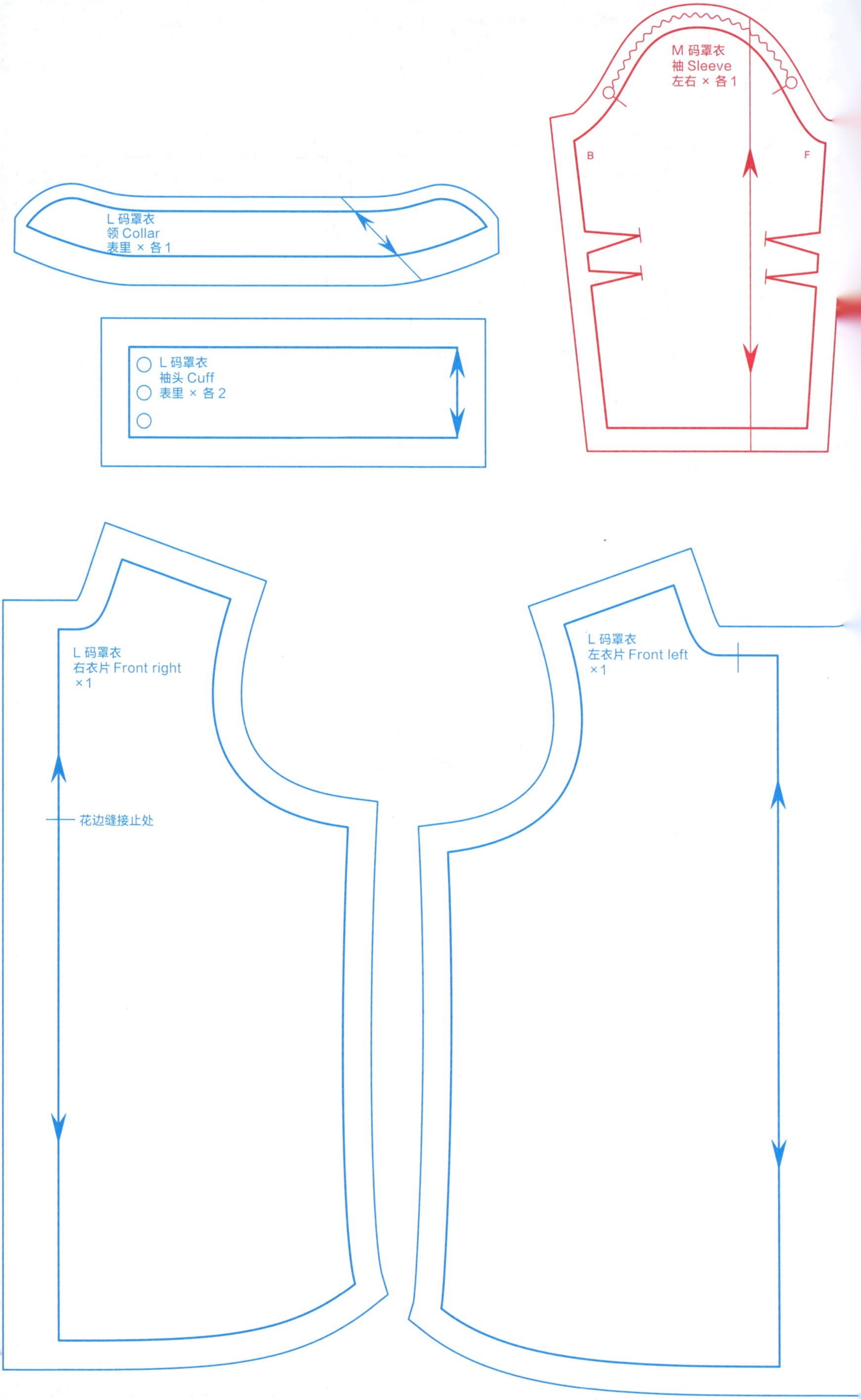
L 码罩衣
领 Collar
表里 × 各 1
L 码罩衣
袖头 Cuff
表里 × 各 2
M 码罩衣
袖 Sleeve
左右 × 各 1
B
F
L 码罩衣
右衣片 Front right
×1
花边缝接止处
L 码罩衣
左衣片 Front left
×1

Skirt
短裙

制作方法：p. 62

右侧开衩止处
L 码短裙
衣片 Skirt
左右 × 各 1

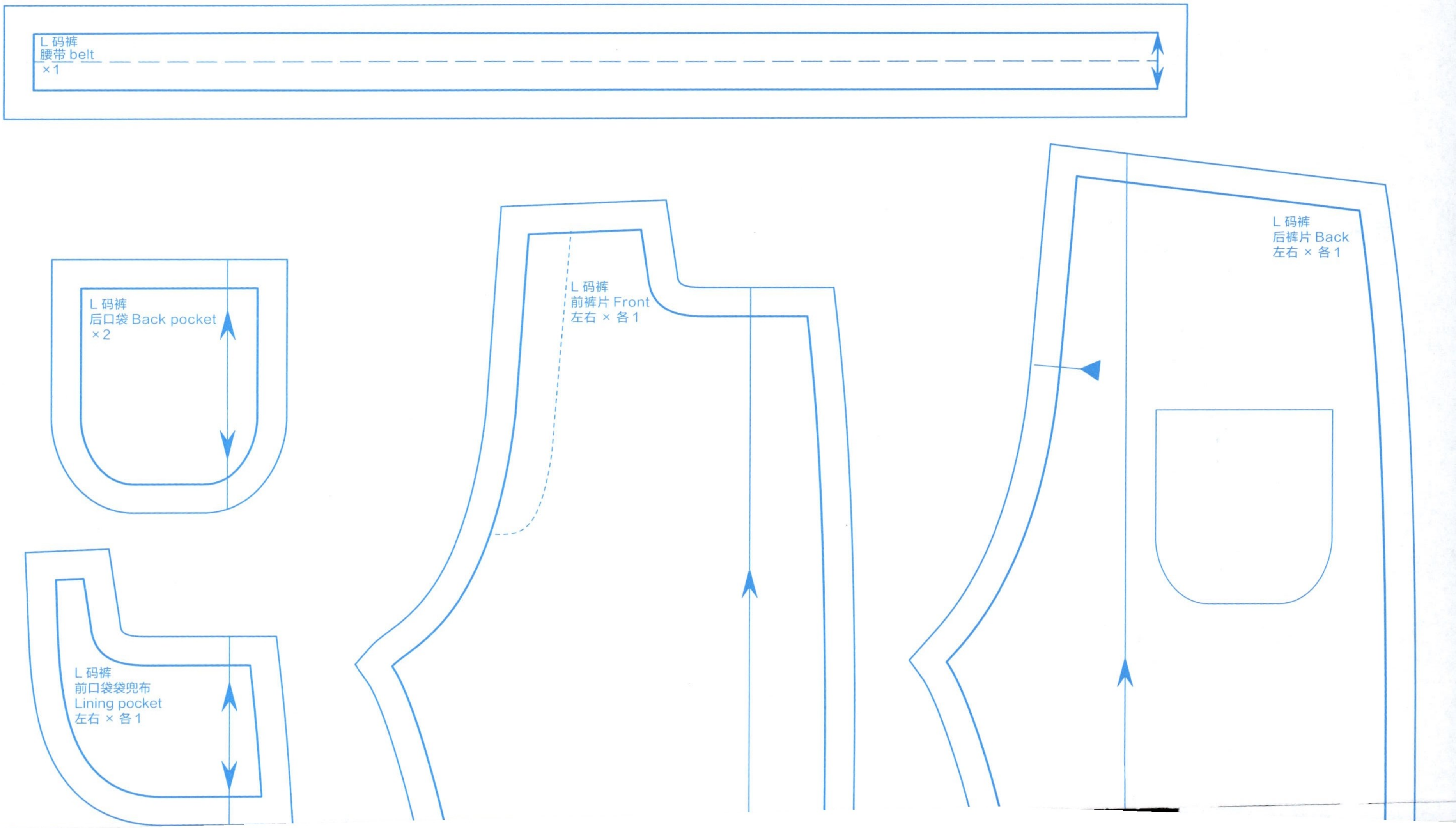

L 码裤
腰带 belt
×1
L 码裤
后口袋 Back pocket
×2
L 码裤
前口袋袋兜布
Lining pocket
左右 × 各 1
L 码裤
前裤片 Front
左右 × 各 1
L 码裤
后裤片 Back
左右 × 各 1

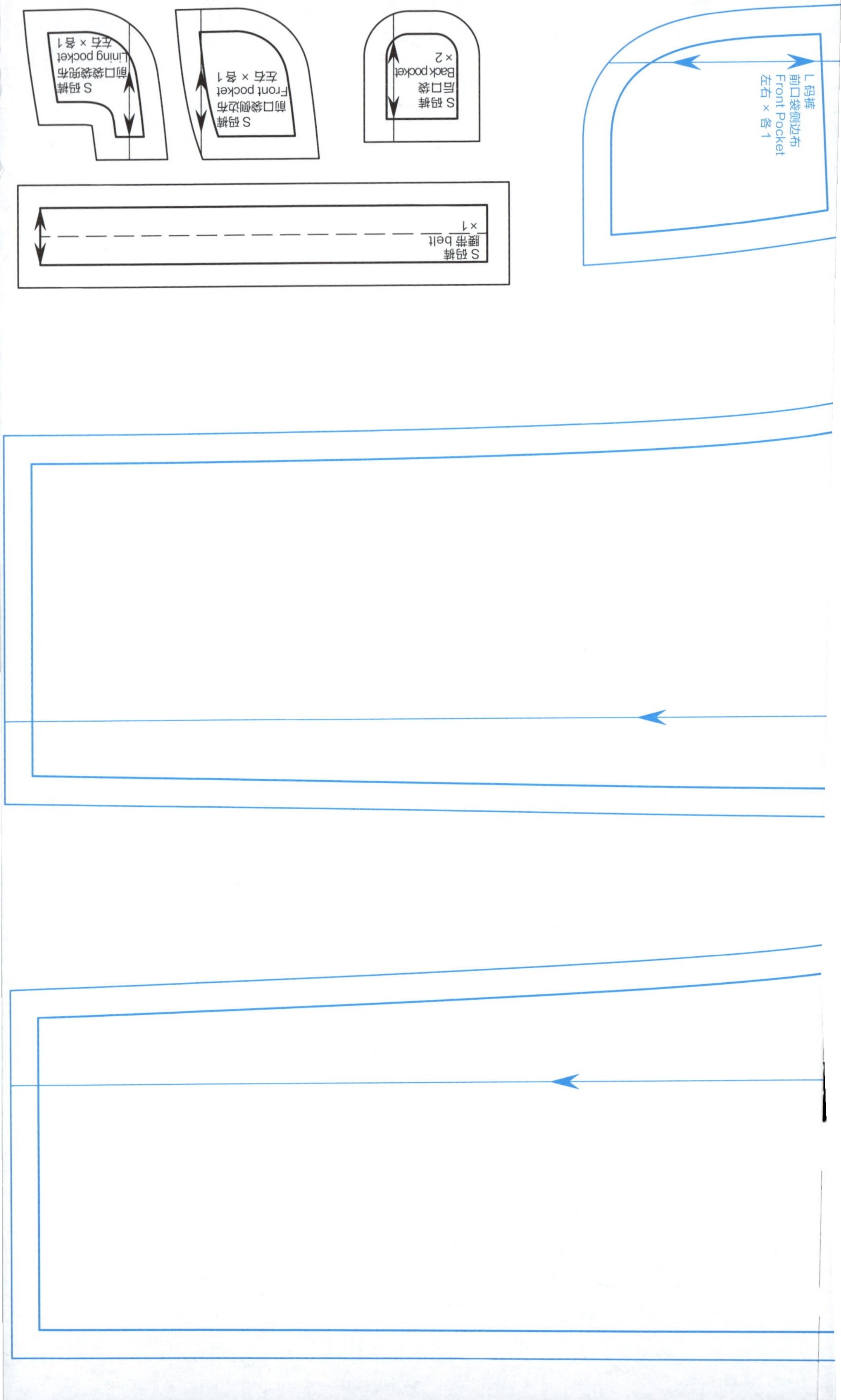
S 码裤
前口袋兜布
Lining pocket
左右 × 各 1
S 码裤
前口袋侧边布
Front pocket
左右 × 各 1
S 码裤
后口袋
Back pocket
× 2
S 码裤
腰带 belt
× 1
L 码裤
前口袋侧边布
Front Pocket
左右 × 各 1

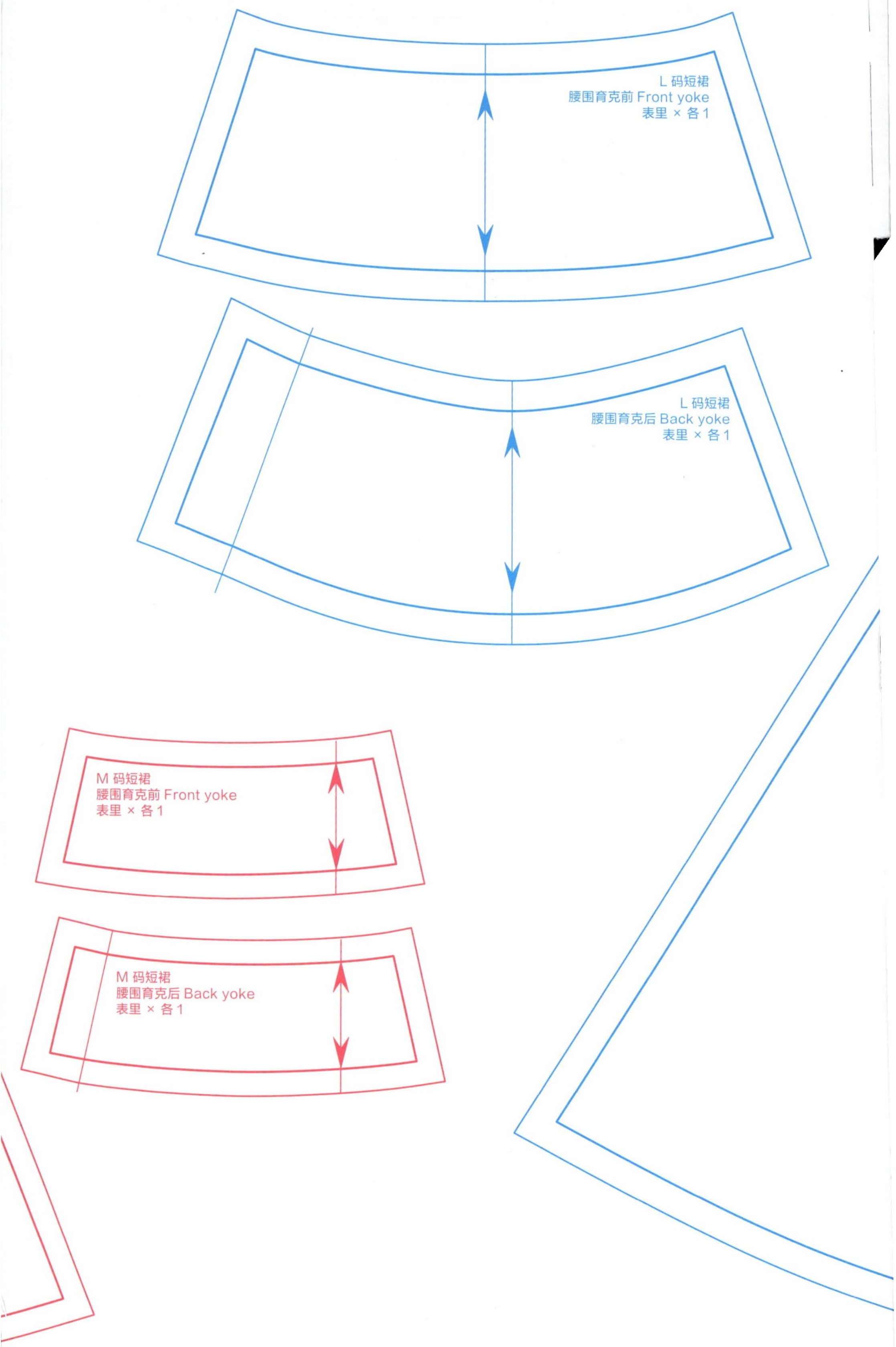
L 码短裙
腰围育克前 Front yoke
表里 × 各 1
L 码短裙
腰围育克后 Back yoke
表里 × 各 1
M 码短裙
腰围育克前 Front yoke
表里 × 各 1
M 码短裙
腰围育克后 Back yoke
表里 × 各 1

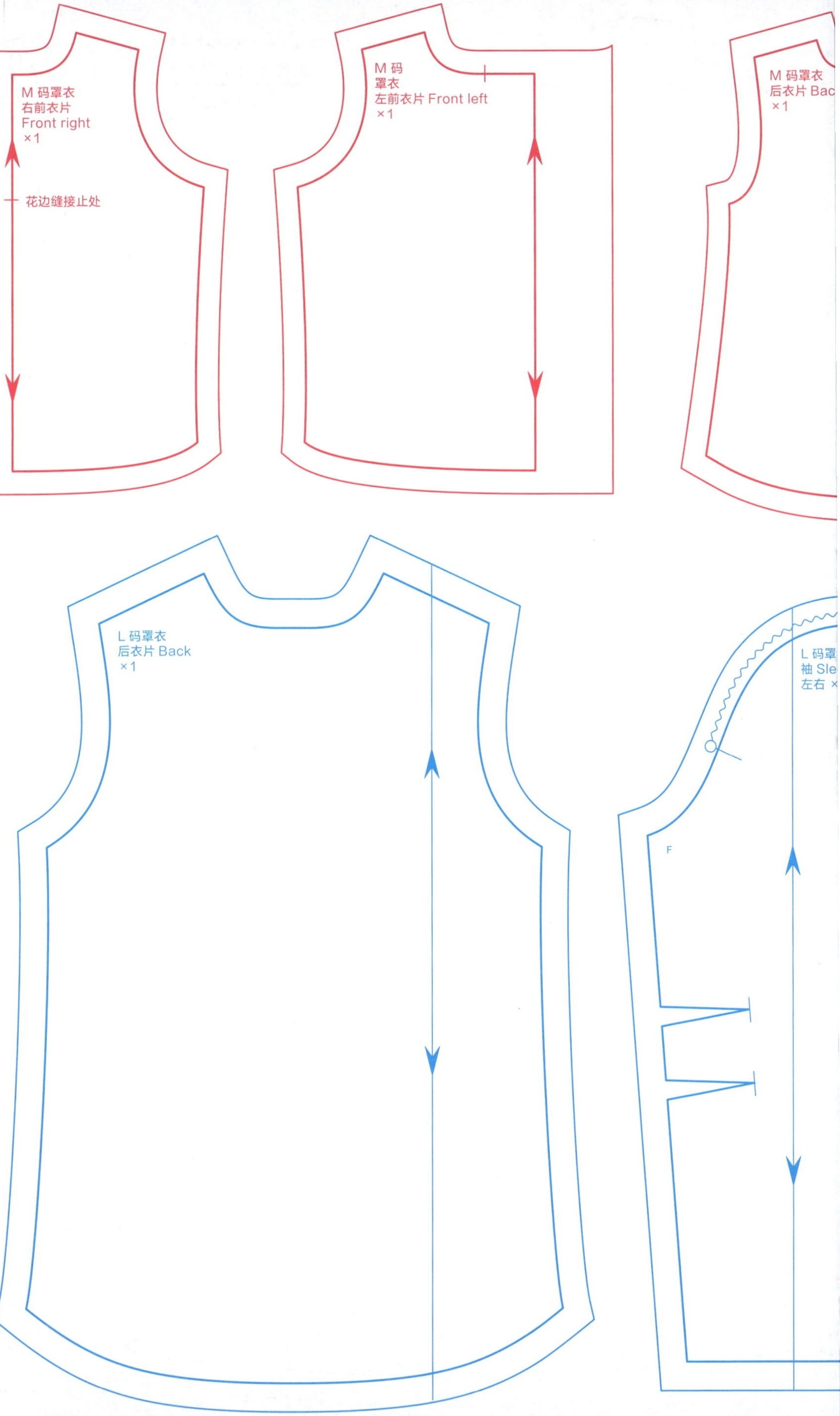

M 码罩衣
右前衣片
Front right
×1
花边缝接止处
M 码
罩衣
左前衣片 Front left
×1
M 码罩衣
后衣片 Bac
×1
L 码罩衣
后衣片 Back
×1
L 码罩
袖 Sle
左右 ×
F

Peter Pan Collar Dress

彼得潘领连衣裙

制作方法：p.50

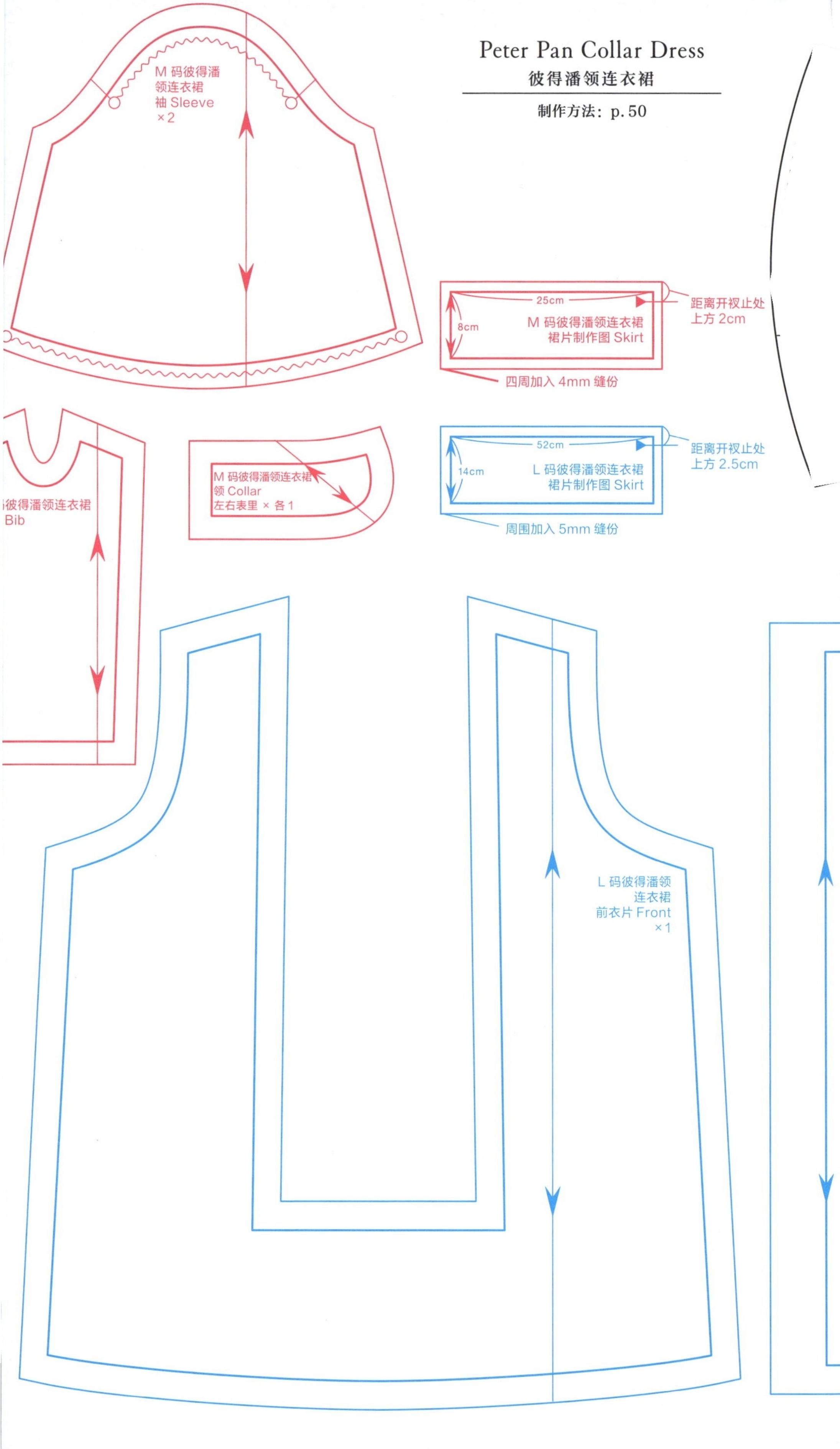
M 码彼得潘
领连衣裙
袖 Sleeve
×2
25cm
8cm
M 码彼得潘领连衣裙
裙片制作图 Skirt
距离开衩止处
上方 2cm
四周加入 4mm 缝份
彼得潘领连衣裙
Bib
M 码彼得潘领连衣裙
领 Collar
左右表里 × 各 1
52cm
14cm
L 码彼得潘领连衣裙
裙片制作图 Skirt
距离开衩止处
上方 2.5cm
周围加入 5mm 缝份
L 码彼得潘领
连衣裙
前衣片 Front
×1

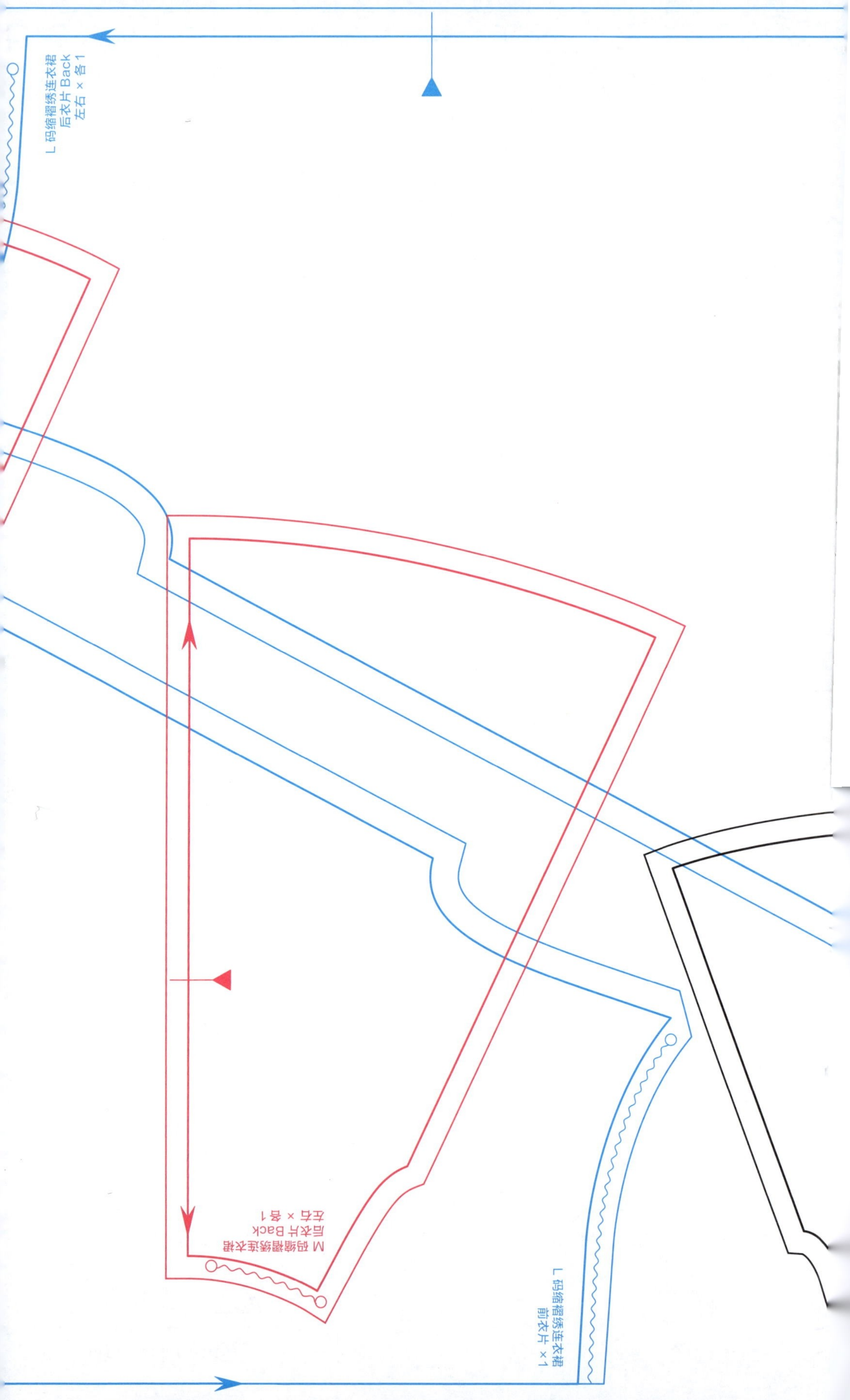
L 码缩褶绣连衣裙
后衣片 Back
左右 × 各1
M 码缩褶绣连衣裙
后衣片 Back
左右 × 各1
L 码缩褶绣连衣裙
前衣片 ×1

Sarrouel Pants
哈伦裤

制作方法：p. 38

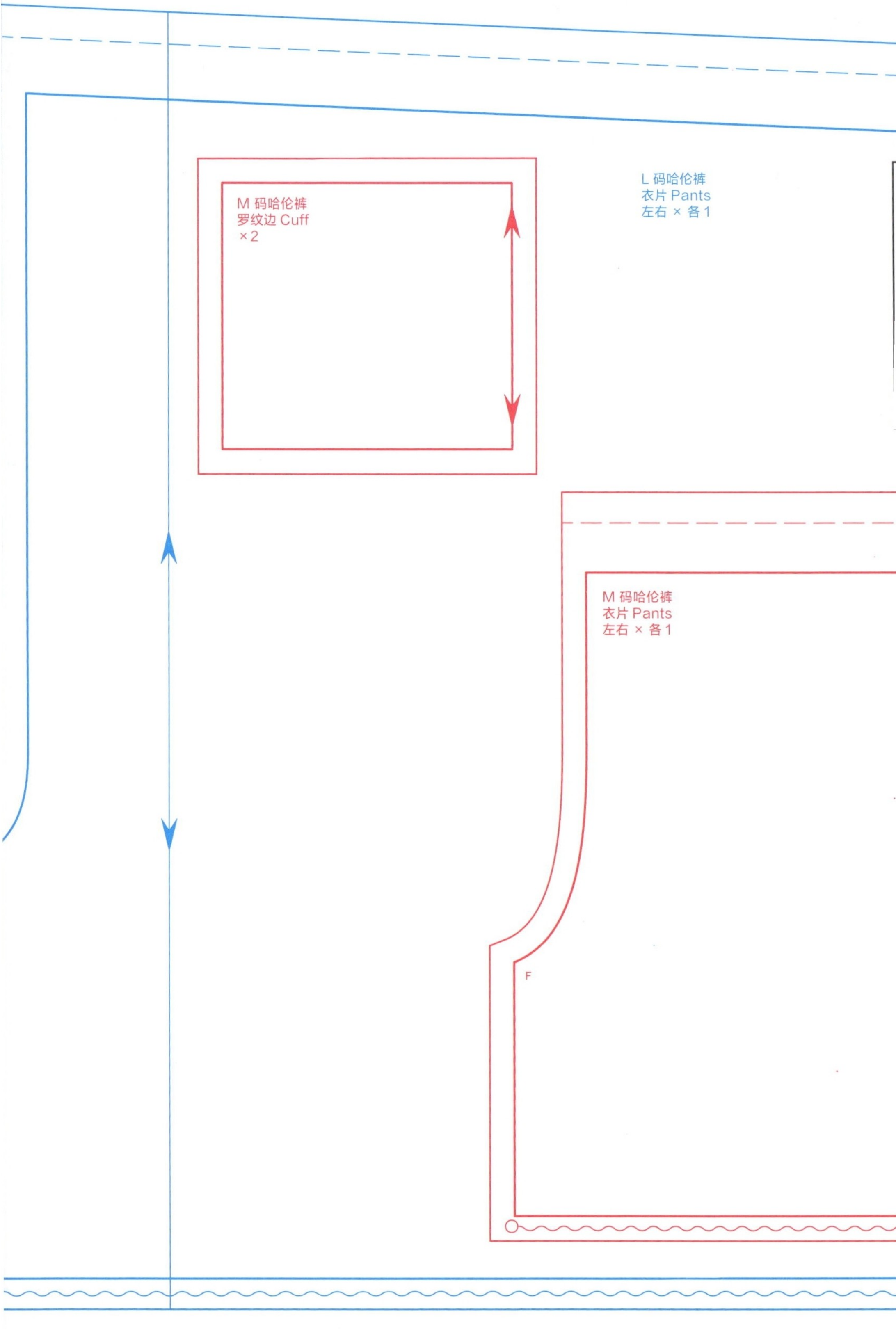
M 码哈伦裤
罗纹边 Cuff
×2
L 码哈伦裤
衣片 Pants
左右 × 各 1
M 码哈伦裤
衣片 Pants
左右 × 各 1
F

Embroidered Smock Dress
缩褶绣连衣裙

制作方法：p. 42

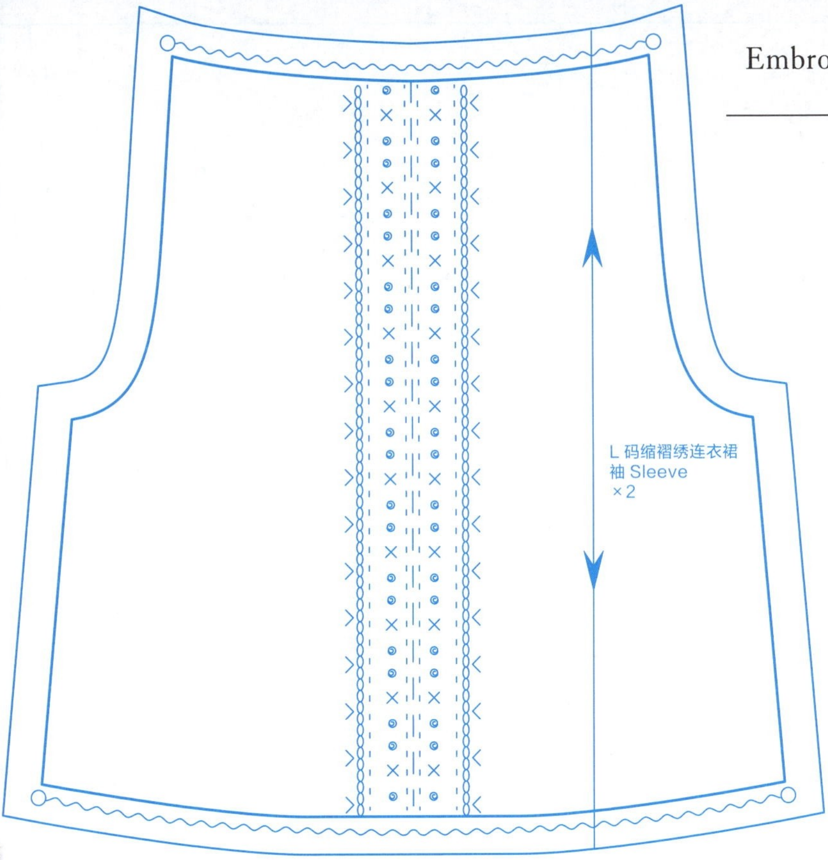
L 码缩褶绣连衣裙
袖 Sleeve
×2

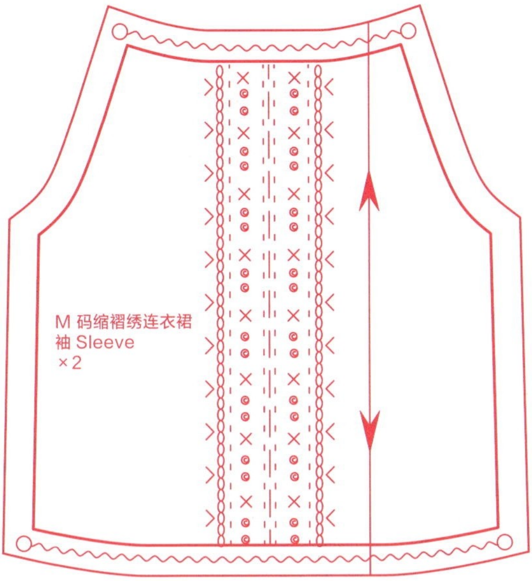
M 码缩褶绣连衣裙
袖 Sleeve
×2

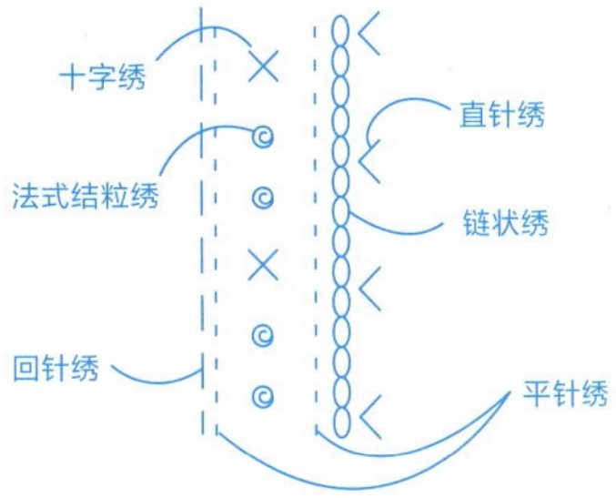
十字绣
法式结粒绣
回针绣
直针绣
链状绣
平针绣

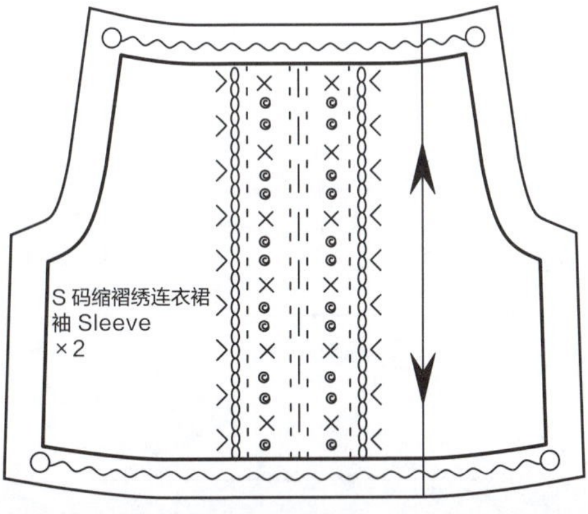
S 码缩褶绣连衣裙
袖 Sleeve
×2

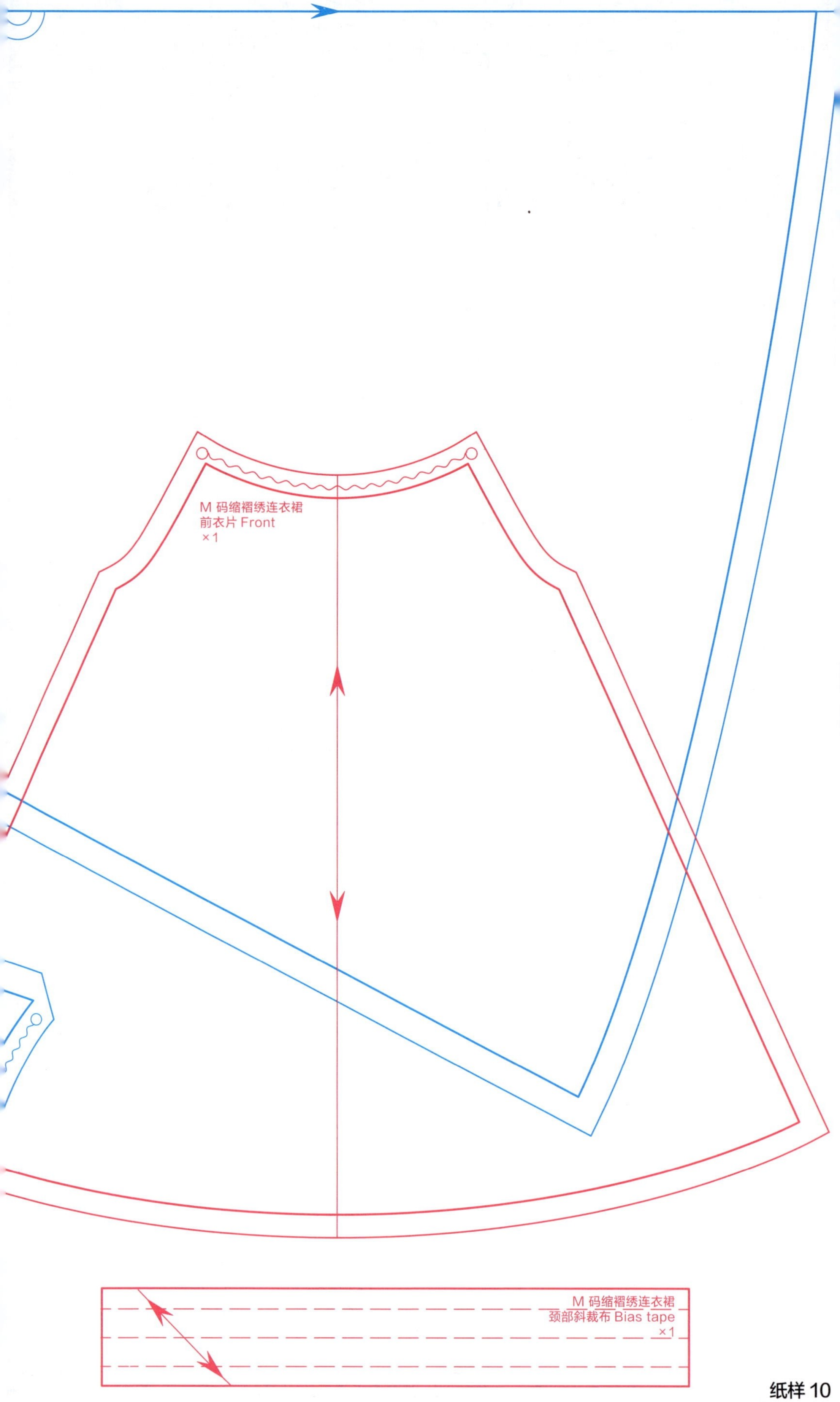
M 码缩褶绣连衣裙
前衣片 Front
×1
M 码缩褶绣连衣裙
颈部斜裁布 Bias tape
×1

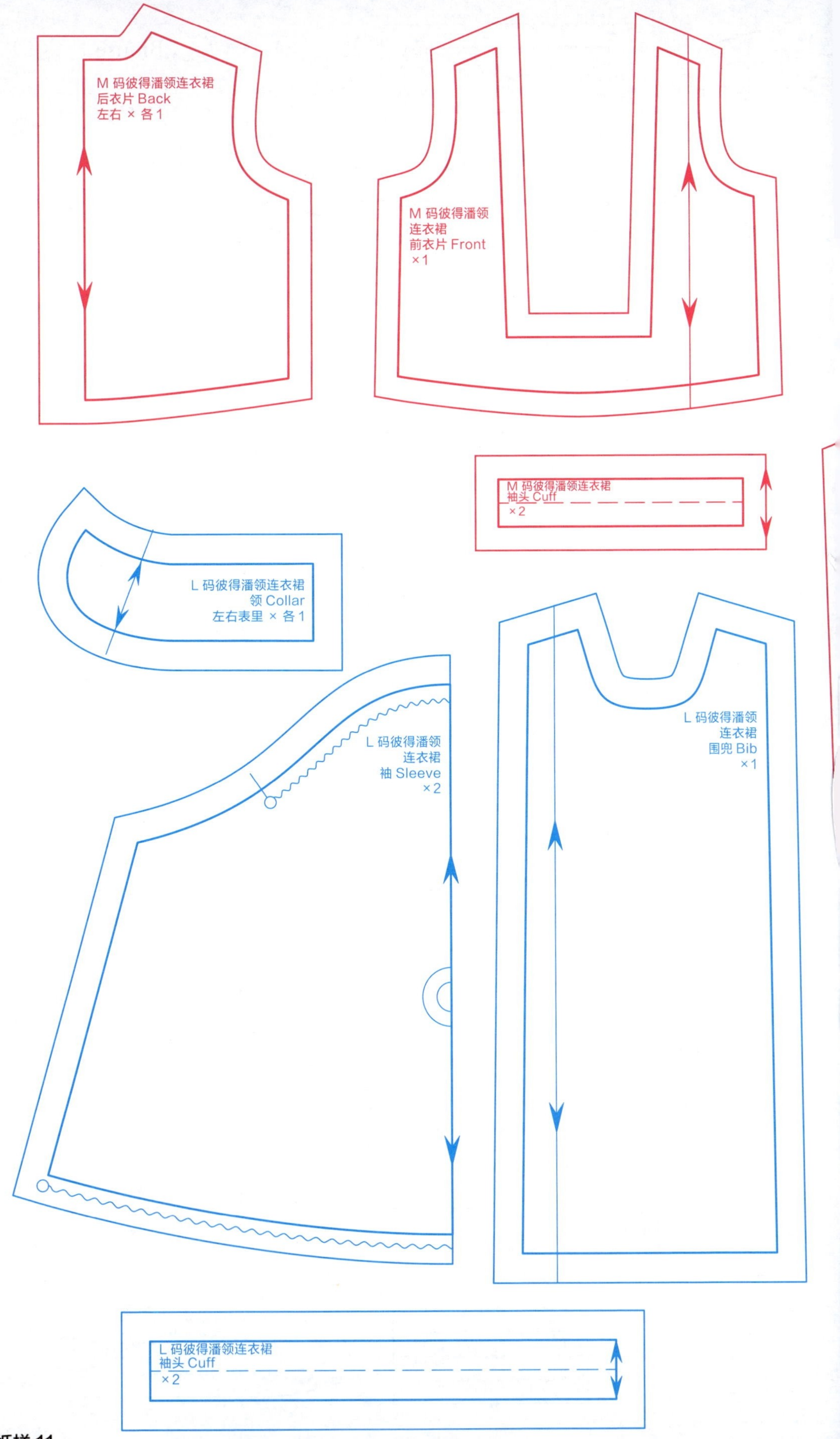
M 码彼得潘领连衣裙
后衣片 Back
左右 × 各 1
M 码彼得潘领
连衣裙
前衣片 Front
×1
M 码彼得潘领连衣裙
袖头 Cuff
×2
L 码彼得潘领连衣裙
领 Collar
左右表里 × 各 1
L 码彼得潘领
连衣裙
袖 Sleeve
×2
L 码彼得潘领
连衣裙
围兜 Bib
×1
L 码彼得潘领连衣裙
袖头 Cuff
×2

Blouse
罩衣

制作方法：p. 56

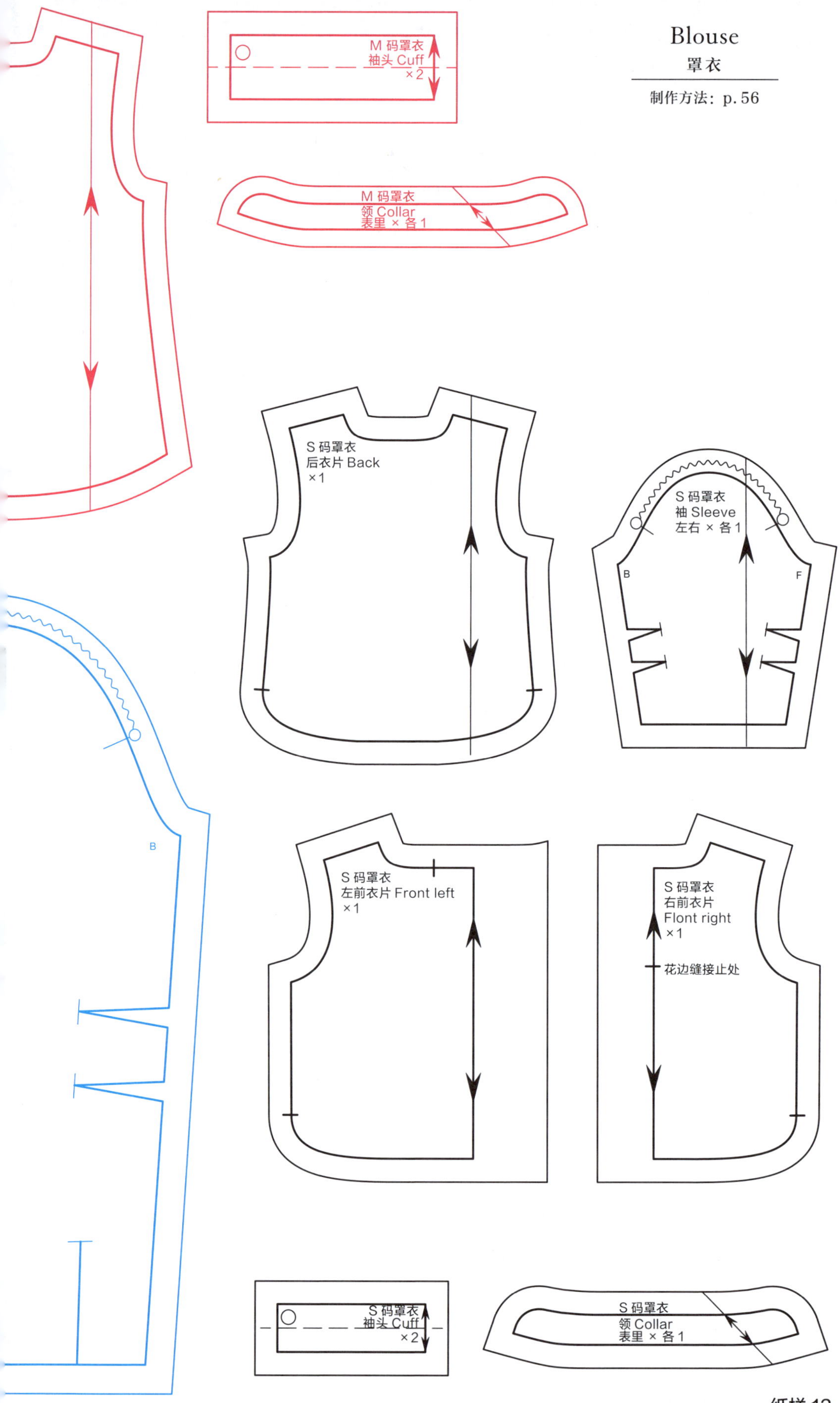
M 码罩衣
袖头 Cuff
×2
M 码罩衣
领 Collar
表里 × 各 1
S 码罩衣
后衣片 Back
×1
S 码罩衣
袖 Sleeve
左右 × 各 1
B
F
B
S 码罩衣
左前衣片 Front left
×1
S 码罩衣
右前衣片
Flont right
×1
花边缝接止处
S 码罩衣
袖头 Cuff
×2
S 码罩衣
领 Collar
表里 × 各 1

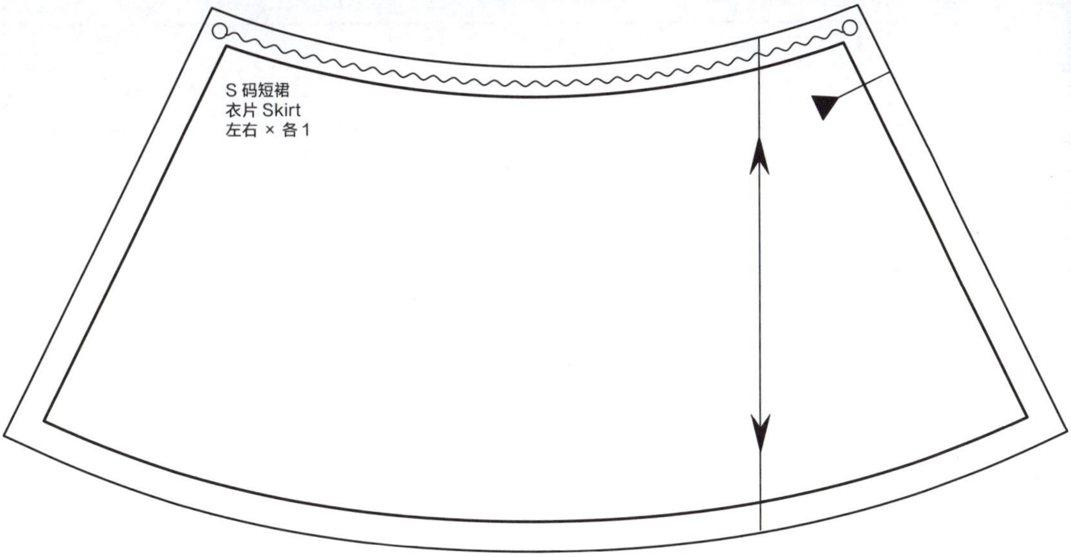
S 码短裙
衣片 Skirt
左右 × 各1

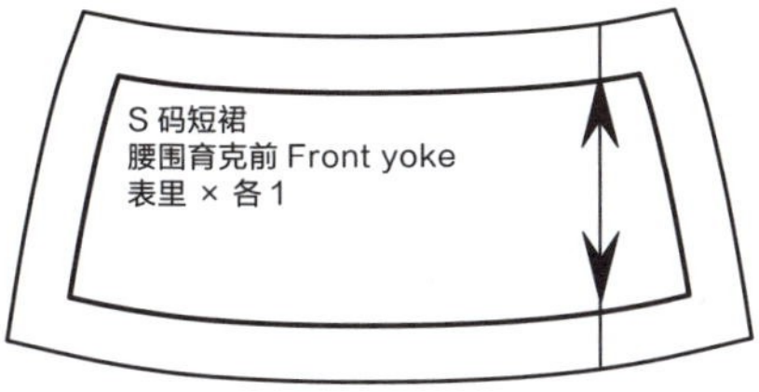
S 码短裙
腰围育克前 Front yoke
表里 × 各1

S 码短裙
腰围育克后 Back yoke
表里 × 各1

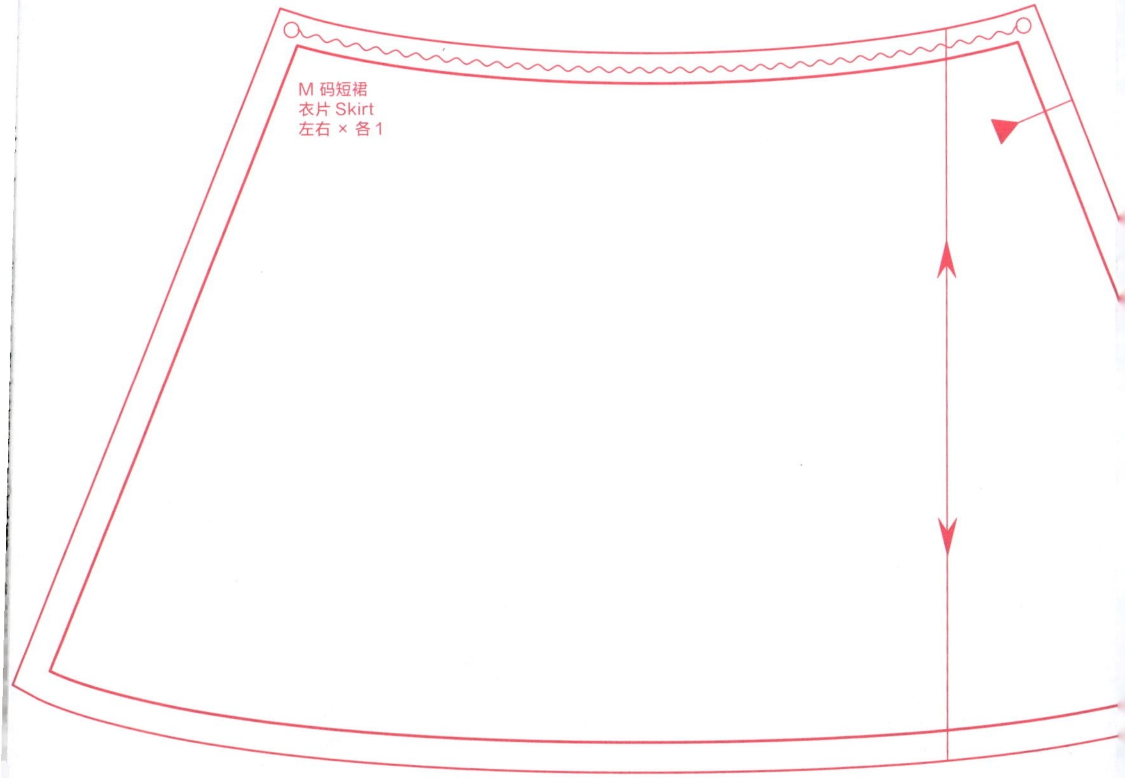
M 码短裙
衣片 Skirt
左右 × 各1

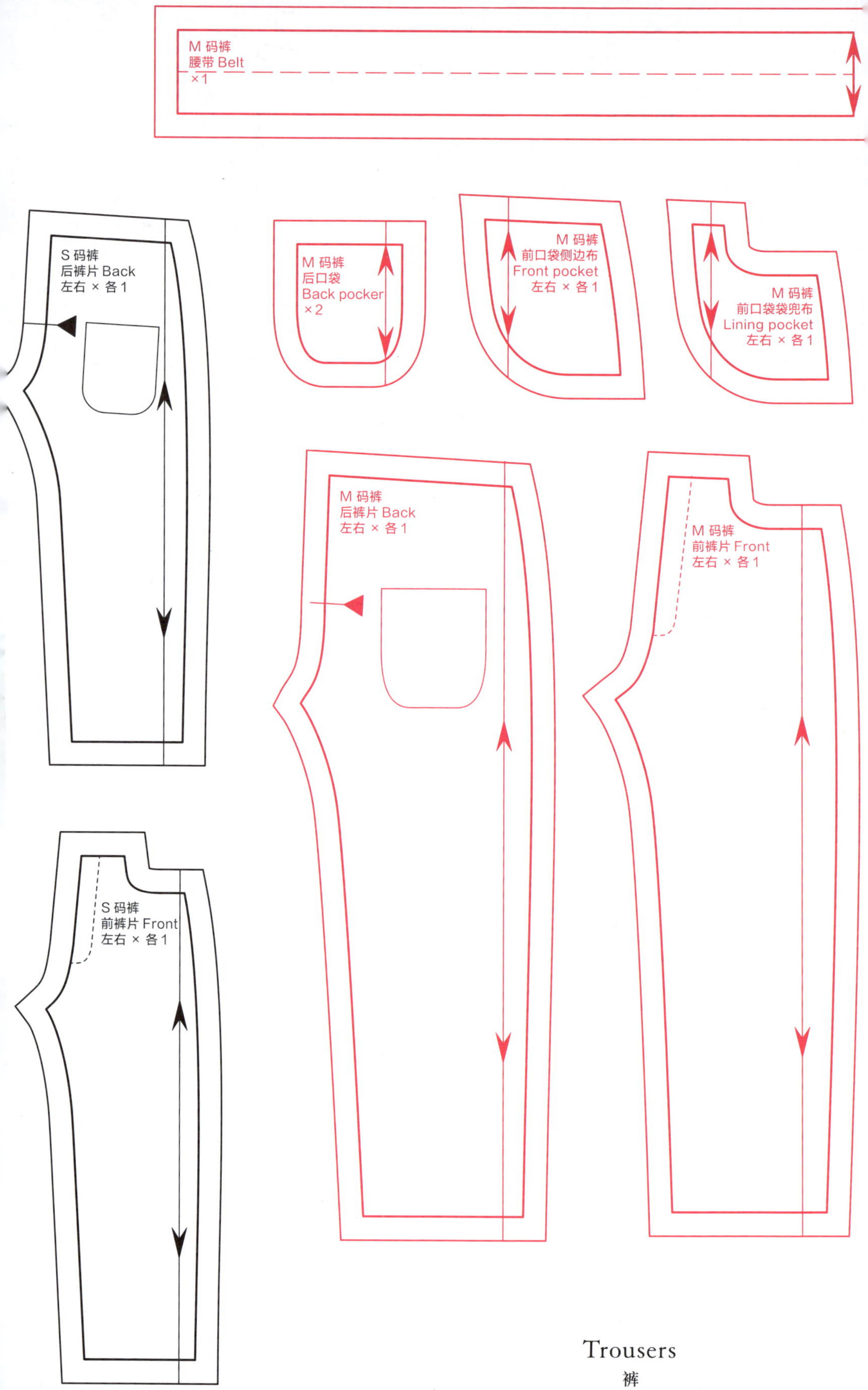

M 码裤
腰带 Belt
×1
S 码裤
后裤片 Back
左右 × 各 1
M 码裤
后口袋
Back pocker
×2
M 码裤
前口袋侧边布
Front pocket
左右 × 各 1
M 码裤
前口袋袋兜布
Lining pocket
左右 × 各 1
M 码裤
后裤片 Back
左右 × 各 1
M 码裤
前裤片 Front
左右 × 各 1
S 码裤
前裤片 Front
左右 × 各 1

Trousers

裤

制作方法：p. 70